D1462264

STATISTICAL METHOD
FROM THE VIEWPOINT OF QUALITY CONTROL

by

WALTER A. SHEWHART

Edited and with a New Foreword by

W. EDWARDS DEMING

DOVER PUBLICATIONS, INC.
New York

Published in Canada by General Publishing Company, Ltd., 30 Lesmill Road, Don Mills, Toronto, Ontario.
Published in the United Kingdom by Constable and Company, Ltd.

This Dover edition, first published in 1986, is an unabridged republication of the work originally published by the Graduate School of the Department of Agriculture, Washington, D.C., in 1939. The original foreword written by W. Edwards Deming has been omitted, and a new foreword has been written by him especially for the Dover edition.

Manufactured in the United States of America
Dover Publications, Inc., 31 East 2nd Street, Mineola, N.Y. 11501

Library of Congress Cataloging-in-Publication Data

Shewhart, Walter A. (Walter Andrew), 1891–1967.
Statistical method from the viewpoint of quality control.

Reprint. Originally published: Washington, D.C.: Graduate School of the Department of Agriculture, 1939.
1. Quality control—Statistical methods. I. Deming, W. Edwards (William Edwards), 1900– . II. Title.
TS156.S468 1986 658.5′62 86-16567
ISBN 0-486-65232-7 (pbk.)

FOREWORD
by W. Edwards Deming

The present book gave Dr. Shewhart an opportunity to elaborate on a number of principles that he had already stated and applied in his great book *The Economic Control of Manufactured Product* (Van Nostrand, 1931) and in a number of papers. The origin of the present book lies in four lectures delivered at my invitation in Washington in 1938 under the auspices of the Graduate School of the Department of Agriculture.

The important principles explained here include Dr. Shewhart's rules for presentation of data (pp. 86–92). Any conclusion or statement, if it is to have use for science or industry, must add to the degree of belief for rational prediction. The reader may reflect on the fact that the only reason to carry out a test is to improve a process, to improve the quality and quantity of the next run or of next year's crop. Important questions in science and industry are how and under what conditions observations may contribute to a rational decision to change or not to change a process to accomplish improvements. A record of observations must accordingly contain all the information that anyone might need in order to make his own prediction.

This information will include not merely numerical data, but also, (for example): the names of observers, the type of apparatus or measurement system used, a description of materials used, the temperature and humidity, a description of efforts taken to reduce error, the side effects and other external factors that in the judgment of the expert in the subject matter may be helpful for use of the results.

Omission from data (as of a test or a run of production) of information on the order of the observations may well bury, for purposes of prediction (i.e., for planning), nearly all the information that there is in the test. Any symmetric function loses the information that is contained in the order of observation. Thus, the mean, standard deviation—in fact any moment—is in most applications inefficient, as it causes the loss of all information that is contained in the order of observation. A distribution is another example of a symmetric function. Original data plotted in order of production may provide much more information than is contained in the distribution.

Dr. Shewhart was well aware that the statistician's levels of significance furnish no measure of belief in a prediction. Probability has use; tests of significance do not.

There is no true value of anything. There is, instead, a figure that is produced by application of a master or ideal method of counting or of measurement. This figure may be accepted as a standard until the method of measurement is supplanted by experts in the subject matter with some other method and some other figure. There is no true value of the speed of light; no true value of the number of inhabitants within the boundaries of (e.g.) Detroit. A count of the number of inhabitants of Detroit is dependent upon the application of arbitrary rules for carrying out the count. Repetition of an experiment or of a count will exhibit variation. Change in the method of measuring the speed of light produces a new result.

All this has been known for generations. What Dr. Shewhart demonstrates in this book is the importance of these principles in science and industry, whether it be manufacturing or service, including government service. The requirements of industry are more exacting than the requirements of pure science.

As with many contributors to science, literature, and the arts, Dr. Shewhart is best known for the least of his contributions—control charts. Control charts alone would be sufficient for eternal fame (even though, because of poor understanding of teachers and books, many students' applications seen in practice are faulty and may be doing more harm than good). The fact is that some of the greatest contributions from control charts lie in areas that are only partially explored so far, such as applications to supervision, management, and systems of measurement, including the standardization and use of instruments, test panels, and standard samples of chemicals and compounds.

The great contribution of control charts is to separate variation by rational methods into two sources: (1) the system itself ("chance causes," Dr. Shewhart called them), the responsibility of management; and (2) assignable causes, called by Deming "special causes," specific to some ephemeral event that can usually be discovered to the satisfaction of the expert on the job, and removed. A process is in *statistical control* when it is no longer afflicted with special causes. The performance of a process that is in statistical control is predictable.

A process has no measurable capability unless it is in statistical control. An instrument has no ascertainable precision unless observations made with it show statistical control. Results obtained by two instruments cannot be usefully compared unless the two instruments are in statistical control. Statistical control is ephemeral; there must be a running record for judging whether the state of statistical control still exists.

Every observation, numerical or otherwise, is subject to variation. Moreover, there is useful information in variation. The closest approach possible to a numerical evaluation of any so-called physical content, to any count, or to any characteristics of a process is a result that emanates from a system of measurement that shows evidence that it is in statistical control.

The expert in the subject matter holds the responsibility for the use of the data from a test.

Another half-century may pass before the full spectrum of Dr. Shewhart's contributions has been revealed in liberal education, science, and industry.

W. E. D.

WASHINGTON
June 1986

The application of statistical methods in mass production makes possible the most efficient use of raw materials and manufacturing processes, effects economies in production, and makes possible the highest economic standards of quality for the manufactured goods used by all of us. The story of the application, however, is of much broader interest. The economic control of quality of manufactured goods is perhaps the simplest type of *scientific* control. Recent studies in this field throw light on such broad questions as: What is the fundamental role of statistical method in such control? How far can man go in controlling his physical environment? How does this depend upon the human factor of intelligence and how upon the element of chance?

PREFACE

Statistical methods of *research* have been highly developed in the field of agriculture. Similarly, statistical methods of *control* have been developed by industry for the purpose of attaining economic control of quality of product in mass production. It is reasonable to expect that much is to be gained by correlating so far as possible the development of these two kinds of statistical technique. In the hope of helping to effect this correlation, it was with pleasure that I accepted the invitation to give a series of four lectures on statistical method from the viewpoint of quality control before the Graduate School of the Department of Agriculture. The subject matter of these lectures is limited to an exposition of some of the elementary but fundamental principles and techniques basic to the efficient use of the statistical method in the attainment of a state of statistical control, the establishment of tolerance limits, the presentation of data, and the specification of accuracy and precision. I am indebted to many, and in particular to Dr. W. Edwards Deming, for the helpful criticisms and stimulating questions brought out in the discussion periods following the lectures and in private conferences.

In preparing these lectures for publication, it has been a pleasure and a privilege to have the wholehearted cooperation of the editor, Dr. Deming, who has contributed many helpful suggestions and has done much to help clarify the text. My colleague, Mr. H. F. Dodge, has given continuing help and advice over the past several years in the development of the material here presented. Miss Miriam Harold has contributed many helpful suggestions at all stages of the work and has for the most part borne the task of accumulating and analyzing the data, drawing the figures, and putting the manuscript in final form. To each of these, I am deeply indebted. For many courtesies extended to me at the time the lectures were given, I am indebted to Dr. A. F. Woods, Director of the Graduate School.

W. A. Shewhart

Bell Telephone Laboratories, Inc.
New York
August 1939

CONTENTS

CHAPTER I—STATISTICAL CONTROL

CHAPTER II—HOW ESTABLISH LIMITS OF VARIABILITY?

CHAPTER IV—THE SPECIFICATION OF ACCURACY AND PRECISION

CHAPTER I

STATISTICAL CONTROL

The possibility of improving the economy of steel to the consumer is therefore largely a matter of improving its uniformity of quality, of fitting steels better for each of the multifarious uses, rather than of any direct lessening of its cost of production.[1]

JOHN JOHNSTON, *Director of Research*
United States Steel Corporation

Introduction. Three steps in quality control. Three senses of statistical control. Broadly speaking, there are three steps in a quality control process: the *specification* of what is wanted, the *production* of things to satisfy the specification, and the *inspection* of the things produced to see if they satisfy the specification. Corresponding to these three steps there are three senses in which statistical control may play an important part in attaining uniformity in the quality of a manufactured product: (*a*) as a concept of a statistical state constituting a limit to which one may hope to go in improving the uniformity of quality; (*b*) as an operation or technique of attaining uniformity; and (*c*) as a judgment. Here we shall be concerned with an exposition of the meaning of statistical control in these three senses and of the role that each sense plays in the theory and technique of economic control. But first we should consider briefly the history of the control of quality up to the time when engineers introduced the statistical control chart technique, which is in itself an *operation of control.*

SOME IMPORTANT HISTORICAL STAGES IN THE CONTROL OF QUALITY

To attain a perspective from which to view recent developments, let us look at fig. 1. That which to a large extent differentiates man from animals is his control of his surroundings and particularly his production and use of tools. Apparently the human race began the fashioning and use of stone tools about a million years ago, as may be inferred from the recent discovery just north of London [2]

Parts fitted 10,000 years ago

[1] "The applications of science to the making and finishing of steel," *Mechanical Engineering*, vol. 57, pp. 79–86, 1935.

[2] This discovery is reported in *Man Rises to Parnassus* by H. F. Osborn (Princeton University Press, 1928). The photograph of the stone implements (fig. 1) of a million years ago has been reproduced by permission from this most interesting book. Those of the implements of 150,000 and 10,000 years ago have been reproduced by permission from the fascinating story told in *Early Steps in Human Progress* by H. J. Peake (J. B. Lippincott, Philadelphia, 1933).

of the crude stone implements shown at the left in fig. 1. Little progress in control seems to have been made, however, until about 10,000 years ago when man began to fit parts together in the fashion evidenced by the holes in the instruments of that day.

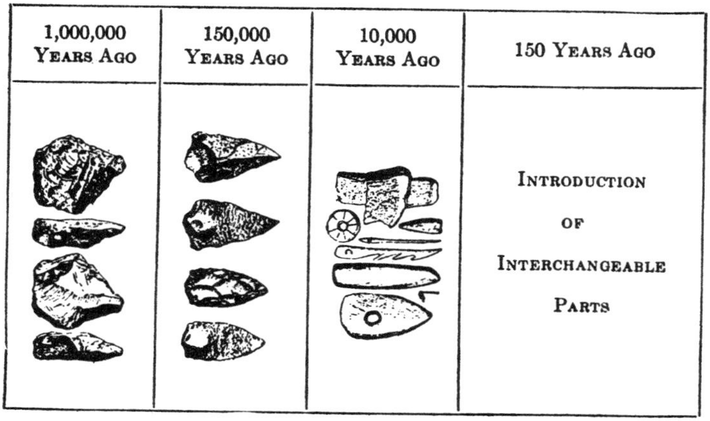

1,000,000 Years Ago	150,000 Years Ago	10,000 Years Ago	150 Years Ago
			Introduction of Interchangeable Parts

Fig. 1

Interchangeable parts—exact, 1787

Throughout this long period, apparently each man made his own tools, such as they were. As far back as 5000 years ago the Egyptians are supposed to have made and used interchangeable bows and arrows to a limited extent, but it was not until about 1787, or about a hundred and fifty years ago, that we had the first real introduction of the concept of interchangeable parts. Only yesterday, as it were, did man first begin to study the technique of mass production!

"Go" tolerance limits, 1840; "go, no-go," 1870

From the viewpoint of ideology, it is significant that this first step was taken under the sway of the concept of an *exact* science, according to which an attempt was made to produce pieceparts to exact dimensions. How strange such a procedure appears to us today, accustomed as we are to the use of tolerances. But as shown in fig. 2, it was not until about 1840 that the concept of a "go" tolerance limit was introduced and not until about 1870 that we find the "go, no-go" tolerance limits.[3]

Why these three steps: "exact," "go," "go, no-go"? The answer is quite simple. Manufacturers soon found that they could not make things

[3] It will be noted that the first six dates shown in fig. 2 are given with question marks —authorities are not in unanimous agreement as to the exact dates. I think, however, that the dates here shown will be admitted by all to be sufficiently close approximations.

exactly alike in respect to a given quality; moreover, it was not necessary that they be exactly alike, and it was too costly to try to make them so. Hence by about 1840 they had eased away from the requirement of exact-

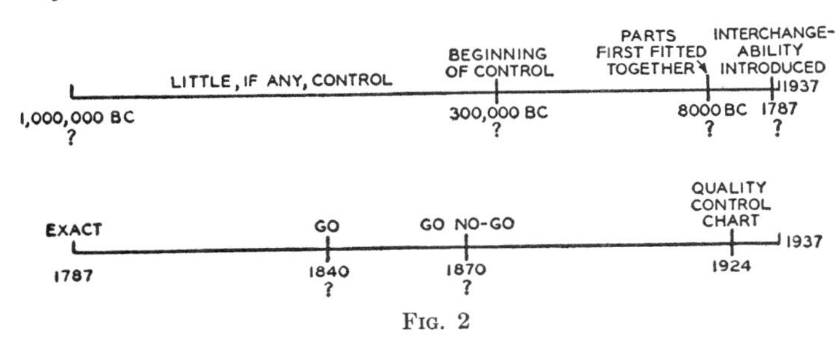

FIG. 2

ness to the "go" tolerance. Let us see how this worked. If we take, for example, a design involving the use of a cylindrical shaft in a bearing, one might insure interchangeability by simply using a suitable "go" plug gauge on the bearing and a suitable "go" ring gauge on the shaft. In this case, the difference between the dimensions of the two "go" gauges gave the minimum clearance. Such a method of gauging, however, did *not fix the maximum clearance.* The production man soon realized that a slack fit between a part and its "go" gauge might result in enough play between the shaft and its bearing to cause rejection, and for this reason he tried to keep the fit between the part and its "go" gauge as close as possible, thus involving some of the same kind of difficulties that had been experienced in trying to make the parts exactly alike. The introduction of the "go, no-go" gauge in 1870 was therefore a big forward step in that it fixed the upper and lower tolerance limits on each fitting part, thus giving the production man more freedom with a resultant reduction in cost. All he had to do was stay within the tolerance limits—he *didn't have to waste time trying to be unnecessarily exact.*

Defective parts; inspection

Though this step was of great importance, something else remained to be done. The limits are necessarily set in such a way that every now and then a piece of product has a quality characteristic falling outside its specified range, and is therefore defective. To junk or modify such pieces adds to the cost of production. But to find the unknown or chance causes of defectives and to try to remove them also costs money. Hence after the introduction of the go, no-go tolerance limits, there remained the problem of trying to reduce the fraction p of defectives to a point where the rate of increase in the cost of control equals the rate of increase in the savings brought about through the decrease in the number of rejected parts.

For example, in the production of the apparatus going into the telephone plant, raw materials are gathered literally from the four corners of the earth. More than 110,000 different kinds of pieceparts are produced. At the various stages of production, inspections are instituted to catch defective parts before they reach the place of final assembly to be thrown out there. At each stage, one must determine the economic minima for the sizes of the piles of defectives thrown out.

This problem of minimizing the percent defective, however, was not the only one that remained to be solved. Tests for many quality characteristics—strength, chemical composition, blowing time of a fuse, and so on—are destructive. Hence not every piece of product can be tested, and engineers must appeal to the use of a sample. But how large a sample should be taken in a given case in order to gain adequate assurance of quality?

Destructive tests; necessity for sampling. How large a sample?

The attempt to solve these two problems gave rise to the introduction of the operation of statistical control involving the use of the quality control chart in 1924, and may therefore be taken as the starting point of the application of statistical technique in the control of the quality of a manufactured product in the sense here considered.

The quality control chart, 1924

Why, you may ask, do we find, some one hundred and fifty years after the start of mass production, this sudden quickening of interest in the application of statistical methods in this field? There are at least two important reasons. First, there was the rapid growth in standardization. Fig. 3 shows the rate of growth in the number of industrial standardization organizations both here and abroad. The first one was organized in Great Britain in 1901. Then beginning in 1917 the realization of the importance of national and even international standards spread rapidly. The fundamental job of these standardizing organizations is to turn out specifications of the aimed-at quality characteristics. But when one comes to write such a specification, he runs into two kinds of problems: (1) *minimizing the number of rejections*, and (2) *minimizing the cost of inspection required to give adequate assurance of quality* in the sense discussed above. Hence the growth in standardization spread the realization of the importance of such problems in industry.

Why after 1900?

Second, there was a more or less radical change in ideology about 1900. We passed from the concept of the exactness of science in 1787, when interchangeability was introduced, to probability and statistical concepts which came into their own in almost every field of science after 1900. Whereas the concept of mass production of 1787 was born of an exact science, the concept underlying the quality control chart technique of 1924 was born of a probable science.

We may for simplicity think of the manufacturer trying to produce a piece of product with a quality characteristic falling within a given tolerance range as being analogous to shooting at a mark. If one of us were shooting at a mark and failed to hit the bull's-eye, and some one asked us why, we

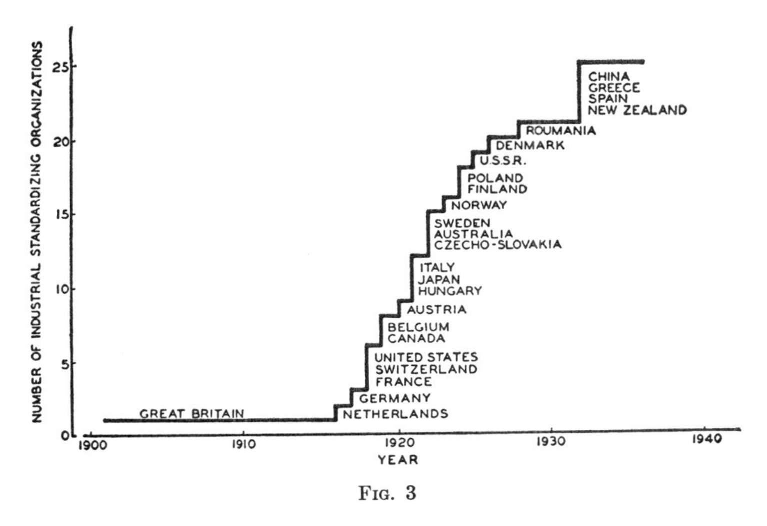

FIG. 3

should likely give as our excuse, CHANCE. Had some one asked the same question of one of our earliest known ancestors, he might have attributed his lack of success to the dictates of fate or to the will of the gods. I am inclined to think that in many ways one of these excuses is just about as good as another. Perhaps we are not much wiser in blaming our failures on chance than our ancestors were in blaming theirs on fate or the gods. But since 1900, the engineer has proved his unwillingness to attribute all such failures to chance. This represents a remarkable change in the ideology that characterizes the developments in the application of statistics in the control of quality.

Developments since 1870. With the introduction of the go, no-go tolerance limits of 1870, it became the more or less generally accepted practice to specify that each important quality characteristic X of a given piece of product should lie within stated limits L_1 and L_2, represented schematically in fig. 4. Such a specification is of the nature of an end requirement on the specified quality characteristic X of a finished piece of product. It provides a basis on which the quality of a given product may be gauged to determine whether or not it meets the specification. From this viewpoint, the process of specification is very simple indeed. Knowing the limits L_1 and L_2 within which it is desirable that a given quality characteristic X should lie, all we need to do is to put these limits in writing as a requirement on the quality of a finished product. With such a specification at hand, the next step is to make the measurements necessary to classify a piece of product as conforming or nonconforming to specification.

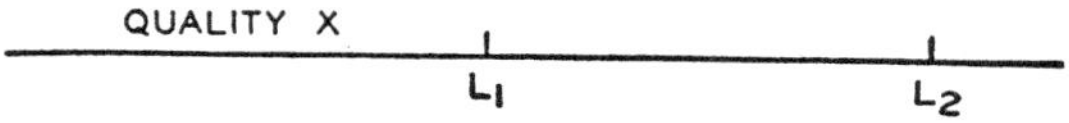

QUALITY X

L_1 L_2

FIG. 4

At this point, however, two problems arise. Suppose that the quality under consideration, the blowing time of a fuse for example, is one that can be determined only by destructive tests. How can one give assurance that the quality of a fuse will meet its specification without destroying the fuse in the process? Or again, even where the quality characteristic can be measured without destruction, there is always a certain fraction p falling outside the tolerance limits. How can we reduce this nonconforming fraction to an economic minimum? A little reflection shows that the simple specification of the go, no-go tolerance limits (p. 3) is not sufficient in such instances from the viewpoint of economy and assurance of quality.

Simple specification of go, no-go tolerance limits often unsatisfactory

As was mentioned at the beginning of this chapter, we shall consider control from the viewpoints of specification, production, and inspection of quality, as is necessary if we are to understand clearly the role played by statistical theory in the economic control of the quality of a manufactured product. To illustrate, suppose we fix our attention on some kind of material, piecepart, or physical object that we wish to produce in large quantities, and let us symbolize the pieces of this product by the letters

$$O_1, O_2, \cdots, O_i, \cdots, O_n, O_{n+1}, \cdots, O_{n+j}, \cdots \qquad (1)$$

presuming that a given process of production may be employed to turn out an indefinitely large number of pieces. We shall soon see that corresponding to the three steps in control there are at least three senses in which the phrase "statistical control" may be used in respect to such an infinite sequence of product.

In the first place, prior to the production of any of the O's, the engineer may propose to attain a sequence of O's that have the property of having been produced under a state of statistical control. In the second place, the engineer, before he starts the production of any specific sequence of objects, is pretty sure to focus his attention on the acts or operations that he wishes to be carried out in the production of the pieces of product. Often, when

The concept of the state of statistical control

the aim is to produce a sequence of objects having a specified quality characteristic within some specified limits, the engineer will refer to the process of production as an *operation of control.* The available scientific and engineering literature, for example, contains many articles discussing "the control of quality" by means of gauges, measuring instruments, and different forms of mechanical technique: much of this literature makes no reference to the use of statistics, though in recent years the actual operations of control have often involved the use of statistical techniques such as, for example, the control chart. In order to distinguish the operation of control in the more general sense from that in which statistical techniques are used for the purpose of attaining a state of statistical control, it is customary to think of the latter as an *operation of statistical control.* That which transforms an operation of control into an operation of statistical control is not simply the use of statistical techniques, but the use of statistical techniques that constitute a *means* of attaining the end characterized here as a state of statistical control. It should be noted that the end desired may be conceived of prior to the production of any sequence of objects symbolized in (1) that have the desired characteristics, and independently of whether any such sequence can be produced. For example, we may conceive of a state of statistical control although we know of no way of attaining such a state in practice. In contrast, before we can describe an operation of statistical control, except to say that it is a means to an end, *we must find by experiment such an operation.*

The operation of control. The operation of statistical control

A requirement regarding control. Let us consider the following specified end *requirement:*

A. The quality of the O's shall be statistically controlled in respect to the quality characteristic X.

As an example, the product might be condensers and the quality characteristic X the capacity; the product might be pieces of steel and X the carbon content; or the product might be any other kind of object with an associated quality characteristic. The natural thing to do is to think of this requirement (A) as expressing a condition that the qualities of a sequence of pieces of product represented by the O's in (1) shall be found to have *when made.* For example, we might, as we shall soon see, interpret this requirement as meaning that the sequence of values of the quality characteristic X belonging to the sequence of objects of (1) shall be random. On the other hand, we might interpret the requirement (A) as implying that the cause system underlying the operation of producing the objects satisfies certain physical requirements. In any case, the requirement itself may be, and usually is, stated prior to the production of any of the O's in (1).

A probable inference regarding control. Now let us contrast the requirement (A) with the following *statement* regarding control:

B. The quality of the O's is statistically controlled in respect to the quality characteristic X.

This is a *judgment* or *probable inference* that the quality of the product actually meets the requirement expressed by (A). Since we are here assuming that the process of manufacture is capable of turning out an indefinitely large number of pieces of product, it follows in practice that the statement (B) implies a prediction about O's not yet made: as a probable inference *it is based on past evidence obtained in the process of making some pieces of product and in testing them.* In other words, it is an inference carried *from the product already made* to that which is to be made in the future. The full meaning of statement (B), as we shall see later, must depend upon a consideration not only of the sense of control implied as a requirement but also as an inference *based upon specific evidence* that this requirement has been met.

It is therefore essential that we examine carefully the three senses of statistical control: *1st*, as a characterization of the state of control; *2d*, as an operation; *3d*, as a judgment. This is necessary if we are to see how the attainment of the economic control of the quality of a manufactured product involves the coordination of effort in the three steps: *specification, production*, and *inspection*, as is depicted graphically in fig. 10, page 45.

The State of Statistical Control

The idea of control involves *action for the purpose of achieving a desired end.* Control in this sense involves both action and a specified end. For example, in the quotation at the head of this chapter we have an expression of the need for controlling the quality of steel to attain the end of greater *uniformity*. The man who is to do the controlling is likely to focus his attention on what he is supposed to do or on what action he is supposed to take in the process of making the steel, whereas the man who uses the steel may be primarily interested in the end result as determined by the quantitative measurements of the quality of the finished product. Hence there are two ways of viewing control in general and statistical control in particular; namely, from the viewpoint of the physical act of production, and from that of the end results as manifested in the uniformity of quality. Correspondingly, there are two ways of conceiving of the state of statistical control; namely, as a physical state describable in physical terms, and as a mathematical state characterized by the quantitative aspects of the end results and describable in mathematical terms and an operation of drawing at random.

Two views of control

Some may prefer to say that there is no mathematical state of control, but instead that there is simply a mathematical *description* of a physical state. This is perfectly satisfactory so far as I see *if* we think of the mathematical description as including an explanation of what the mathematical statistician means by the *operation* of drawing a sample at random and not simply the description of the results that he obtains mathematically. However, in much the same sense, there is no *observable* physical state of control except in descriptive terms characterizing some operation such as drawing a sample with replacement from a bowl, repeating an observation under the same essential conditions, or going as far as one can go in the process of controlling quality by finding and removing causes of variability. To be more exact, therefore, we should perhaps speak of the physical and mathematical *descriptions* of the state of control, but it will simplify matters to speak only of the "physical and mathematical states" in our attempt to relate the physical and mathematical operations.

As a background for our consideration of the two states of statistical control, we shall start with the aim of the engineer to manufacture a product of uniform quality. We shall take this to imply that the quality should be *reproducible within limits,*[4] or that the engineer should be able to predict with minimum error the percentage of the future product that will be turned out by a given process with a quality within specified limits. The engineer desires to reduce the variability in quality to an economic minimum. In other words, he wants

(*a*) a rational method of prediction that is subject to minimum error, and
(*b*) a means of minimizing the variability in the quality of a given product at a given cost of production.

Is it possible to control the production process so that these two wants may be satisfied? If so, how shall the engineer know when the production process is in such a state of control? How can this state be characterized? Shall it be by describing the physical operations that the engineer goes through in producing the product; shall it be in terms of quantitative data obtainable from the product in such a state of control; or shall it be by means of a combination of the two? As a basis for answering such questions, we must consider on the one hand the *physical aspects* of the state of control, and on the other hand the *mathematical aspects* of the quantitative data obtainable under a given state of control.

The physical state of statistical control. The ideal bowl experiment. Let us consider first an idealized experiment. Let us assume that we have

[4] Sometimes the term homogeneous is used instead of the more descriptive phrase "reproducible within limits."

N physically similar chips on each of which is written a number. We place these in a bowl and draw successive samples of n chips one at a time with replacement and thorough mixing.[5] Experience shows that the differences between samples drawn under such conditions are predictable in a probability sense and that there is nothing that we can do to reduce the variability in the complexion of the samples. Hence the physical operation involved in getting such a series of samples constitutes an empirical means of describing a physical state of statistical control.

However, the engineer does not deal with drawings from a bowl. Instead he deals with measurements of one kind or another. Let us assume that it is possible to attain a physical state of statistical control of such measurements. How does the engineer set about attaining such a state? The answer is that in making a series of repetitive measurements of a physical constant or in producing units of the same kind of product, he tries to control all of the causes of variability until he has attained a state where the conditions remain, as he says,[6] "essentially the same."

It may be helpful to note that the concept of a physical state of statistical control as illustrated by the example of drawings from a bowl appears to be much the same as the concept of doing something "physically at random." For example, Neyman [7] says: "There are experiments which, even if carried out repeatedly with utmost care to keep the conditions constant, yield varying results. They are 'random.'" Does this mean that we can rely upon our ability to perceive when conditions are being controlled with the utmost care, and that we shall not go astray by calling such experiments random and acting as though they were random? It seems to me that it is far safer to take some *one* physical operation such as drawing from a bowl as a physical model for an act that may be repeated at random, and then to require that any other repetitive operation believed to be random shall *in addition* produce results similar in certain respects to the results of drawing from a bowl before we act as though the operation in question were random. This seems particularly advisable in the light of my own experience which

[5] See my *Economic Control of Quality* (Van Nostrand, New York, 1931); on p. 164 is a description of a normal bowl, and in Appendix II is a record of 4000 drawings therefrom, together with various calculations on them.

[6] Such a characterization of a physical state of statistical control is subjective, and usually all authorities will not agree in a given case that such a state has been attained. It is true that a subjective judgment is involved in setting up the ideal bowl experiment of the previous paragraph. Experience shows, however, that in the case of the bowl, probability theory is usually applicable, and fluctuations in the complexion of the samples are usually accepted as being at a minimum. My own experience indicates that this situation does not hold, in general, for fluctuations in measurements arising under conditions merely judged to remain essentially the same.

[7] J. Neyman, *Lectures and Conferences in Mathematical Statistics* (The Graduate School, The Department of Agriculture, Washington, 1938), p. 21.

indicates that almost all sets of data taken under "presumably the same essential conditions" fail to satisfy the additional requirement that they be in certain ways like those drawn from a bowl (see chapters II and III).

The concept of a state of statistical control must define in an abstract way *the* physical state of statistical control, and hence something supposedly common to all specific instances. Thus even if we agree that sampling from a bowl constitutes a physical state of statistical control, what is there common about such a physical state and any physical state of statistical control of some production process? The answer appears to be: By their results we shall know them. The only way in which we may hope to define objectively a common characteristic of such states is in terms of certain quantitative aspects of their observable characteristics. But in order to get such a basis of comparison, we must go to mathematics and try to find some abstract way of describing a state of statistical control in terms of characteristics of sequences of numbers that we expect to get by repeating an operation arbitrarily chosen as a random one.

In trying to formulate some of the important characteristics of a useful concept of a mathematical state of statistical control we should keep in mind that the state of statistical control is something presumably to be desired, something to which one may hope to attain; in other words it is an ideal goal. We may conceive of this state prior to the act of attaining it in a given instance and irrespective of whether it can be attained in practice. The delineation of such a concept is a priori and definitive, whereas the application of the concept to a particular given physical state of control is hypothetical. The concept of a state of control is used in this definitive sense in the requirement A (p. 7) that the quality of the product shall be statistically controlled in respect to the quality characteristic X, whereas it appears in the hypothetical sense in the corresponding judgment B (p. 8). In order to be of practical use, the state of statistical control should not be defined solely in terms of either the physical cause system or the results produced by the cause system. Instead, it should be defined in terms of both the perceivable characteristics of a cause system that is capable of producing an infinite sequence, and the quantitative characteristics of the infinite sequence produced by such a cause system.

The mathematical state of statistical control. Let us think of the chance cause system as controlling the variation in quality of a given product in such a way that the expected frequency dp of producing a piece of product with a quality characteristic X lying within the range $X \pm \frac{1}{2}\,dX$ is given by an expression of the form

$$dp = f(X)\,dX \tag{2}$$

where f is a mathematical function.

The mathematician may be inclined to accept equation (2) as defining mathematically a statistical universe representing a statistical state of control in respect to the quality X. However, let us see why equation (2) can not be taken as the complete description of what is here meant by statistical state. Let us consider, for example, a production process turning out an indefinitely large sequence (1) of objects having a certain quality X. Let

$$X_1, X_2, \cdots, X_i, \cdots, X_n, X_{n+1}, \cdots, X_{n+j}, \cdots \qquad (3)$$

represent single measurements on the qualities of such a sequence of objects taken in the order of their production. Can we use equation (2) as a basis for determining whether the sequence (3) arose under statistically controlled conditions? There are, as we shall now see, three reasons why this can not be done.

First, even for the infinite sequence, there is no unique function f to be used as a basis for comparison. Second, equation (2) describes a property of an infinite sequence that is approached as a statistical limit and not a property of a finite portion thereof such as we always have in practice. Third, there is nothing in such a definition that explicitly places any restriction on the *order* in the sequence (3) even though it is essential that this order should be what the mathematician refers to as random.

Need for differentiating between a universe and a statistical state of control

Let us consider in more detail each of these three limitations. Some of the earliest attempts to characterize a state of statistical control were inspired by the belief that there existed a special form of frequency function f and it was early argued that the normal law characterized such a state. When the normal law was found to be inadequate, then generalized functional forms were tried. Today, however, all hopes of finding a unique functional form f are blasted. Even if there did exist such a unique function f, we should still be faced with a second difficulty, namely, that such a function would be descriptive of a property of the whole of the infinite sequence and not of a part of it. In consequence, we should have to take a comparatively large sample before we should be justified in judging whether the degree of fit between a theoretical curve and the observed distribution in a finite portion of the sequence warranted the belief that the sequence was statistically controlled. Worse than that, however, is the fact that the functional form of a distribution is independent of the order of occurrence of the observed values in a sequence and hence is not a criterion of randomness.

In this connection, it will have been noted that in stating equation (2) we have spoken of dp as an expected frequency, and not as a probability as is often done. In general, we may say that a variable X has the frequency function $f(X)$ if the frequency of occurrence of X in any arbitrarily chosen range $\alpha < X < \beta$ is measured by $\int_\alpha^\beta f(X)\,dX$, the frequency function being so defined that the integral between $+\infty$ and $-\infty$ is equal to unity. We may take the concept of frequency as primary and essentially undefined. Often it is said that this integral expresses the probability that X lies between α and β. It should be noted, however, that the frequency expressed by this integral is a property of the infinite sequence as a whole and does not necessarily fix the order in the sequence. On the other hand, to state that the *probability* that X will fall within the interval $\alpha < X < \beta$ is equal to the integral $\int_\alpha^\beta f(X)\,dX$ implies that the variable exhibits what we usually speak of as a random order. *The concept of a mathematical state of statistical control must involve some operationally definite meaning for random order.*

An attempt at defining random order for infinite sequences. What the engineer would like to have, therefore, is an infinite sequence of numbers that would characterize once and for all the order that a statistically controlled state of causes may be expected to give. Let us assume for a moment that the numbers

$$s_1,\ s_2,\ \cdots,\ s_i,\ \cdots,\ s_n,\ s_{n+1},\ \cdots,\ s_{n+j},\ \cdots \tag{4}$$

constitute such a sequence. How then should we compare sequences (3) and (4)? Particularly, how should this be done when we have observed only a finite number n of terms of the infinite sequence (3)? These are questions calling for the cooperation of the mathematical statistician.

First, let us consider briefly the problem of characterizing once and for all a *random* comparison sequence symbolized by (4). We may start with a consideration of the method proposed by von Mises[8] and others. In accord with this proposal, two requirements are placed on an infinite sequence in order that it may be called random. Let p be the fraction of the first n numbers in the infinite sequence (3) lying within any arbitrarily chosen interval $\alpha < X < \beta$. Then the *first requirement* is that the limit

Requirements devised by von Mises; difficulties encountered

$$\lim_{n\to\infty} p = p' \tag{5}$$

[8] For a statement of the requirements here attributed to von Mises, see H. Cramer, *Random Variables and Probability Distributions* (Cambridge, 1937), p. 4. See also S. S. Wilks, *Statistical Inference* (Princeton Mathematical Notes, 1937), pp. 1, 2.

shall exist where p' is a constant. The *second requirement* is that the analogous limit shall exist and have the same value p' for every subsequence that can be formed from (3) according to a specified operation A such that it can always be decided whether the ith observation of (3) should belong to a given subsequence *without knowing the magnitude of this particular observation.*[9] It may be helpful to symbolize this procedure as follows:

The original infinite sequence
$$\left\{ X_1,\ X_2,\ \cdots,\ X_i,\ \cdots,\ X_n,\ X_{n+1},\ \cdots,\ X_{n+j},\ \cdots \right. \tag{3}$$

An infinite number of infinite sequences each derived by rearranging (3) according to some specified operation A
$$\left\{ \begin{array}{l} X_{11},\ X_{12},\ \cdots,\ X_{1i},\ \cdots,\ X_{1n},\ X_{1,\,n+1},\ \cdots,\ X_{1,\,n+j},\ \cdots \\ X_{21},\ X_{22},\ \cdots,\ X_{2i},\ \cdots,\ X_{2n},\ X_{2,\,n+1},\ \cdots,\ X_{2,\,n+j},\ \cdots \\ \cdot \\ \cdot \\ \cdot \\ X_{k1},\ X_{k2},\ \cdots,\ X_{ki},\ \cdots,\ X_{kn},\ X_{k,\,n+1},\ \cdots,\ X_{k,\,n+j},\ \cdots \\ \cdot \\ \cdot \\ \cdot \end{array} \right. \tag{3a}$$

It is to be understood that every number X_{ij} in the infinite set of sequences (3*a*) is a member of (3), and that every member of (3) is to be used once and only once in any one sequence of (3*a*)

In general, then, the test for randomness of an infinite sequence (3) becomes one of determining whether or not the original sequence belongs to the class created by the operation A, as fixed by the two requirements just stated.

At least three difficulties arise in trying to use this concept of randomness in quality control work. In the first place, it is recognized[10] that ignorance of the magnitude of X is not a good criterion of independence in selection. For example, Kendall and Smith argue that there is no such thing as a random selection from a universe considered apart from the universe whose members are being selected. In the second place, there is no available practical means of comparing sets of infinite sequences in the way proposed. In the third place, we never have an observed infinite sequence (3) to start with. What the practical man wants is a method for determining whether or not a finite sequence consisting, let us say, of the first n terms of (3) is random.

[9] *Cf.* H. Cramer, *Random Variables and Probability Distributions* (Cambridge Univ. Press, 1937), p. 4.

[10] For an interesting discussion of randomness from a viewpoint much the same as here presented, see the article by M. G. Kendall and B. Babington Smith, "Randomness and random sampling numbers," *Jour. Roy. Stat. Soc.*, vol. ci, pp. 147–166, 1938.

An attempt at defining random order for finite sequences. Let us undertake to determine in what way it is meaningful in an operationally verifiable sense to ask whether an observed sequence of n terms is random. If we choose, as I have done above, to consider the operation of drawing from a bowl random, we may theoretically obtain an infinite class of finite sequences composed of these same n numbers in the following simple way. Let us write the n numbers on as many symmetrical chips, put the chips in a bowl and mix them thoroughly. Then let us draw the numbers out one at a time without replacement and record them in the sequence drawn. By repeating this process indefinitely, we get an infinite set (3b) of finite sequences of n numbers each with which to compare the original sequence.

$$\begin{array}{l} \text{Infinite set of finite sequences,} \\ \text{each being one of the } n! \text{ possible} \\ \text{orders in which the } n \text{ chips can} \\ \text{be drawn from the bowl} \end{array} \left\{ \begin{array}{l} X_{11},\ X_{12},\ X_{13},\ \cdots,\ X_{1n} \\ X_{21},\ X_{22},\ X_{23},\ \cdots,\ X_{2n} \\ X_{31},\ X_{32},\ X_{33},\ \cdots,\ X_{3n} \\ \cdot \\ \cdot \\ \cdot \\ X_{k1},\ X_{k2},\ X_{k3},\ \cdots,\ X_{kn} \\ \cdot \\ \cdot \\ \cdot \end{array} \right. \qquad (3b)$$

It is to be understood that any number X_{ij} in this infinite set of sequences is some one of the n numbers drawn from the bowl.

Now since only $n!$ different sequences are possible with n chips, by the time we have drawn $n! + 1$ sequences, some one of the possible $n!$ orders must have been repeated at least once. It is usual to assume that in the infinite set of sequences, all orders occur with equal frequency. On this basis the order of the original observed sequence is one of the $n!$ possible orders and it is neither more nor less likely to occur in the infinite set of comparison sequences than any other order. For some such reason it has long been argued cogently by many that we can not hope to define a random sequence in terms of the properties of that sequence.

Suppose, however, that a scientist or an engineer were to observe a sequence of n values of X in which, let us say, each succeeding value of X is either equal to or greater than the preceding one, as for example, in the monotonic sequence

$$X_1 \leq X_2 \leq X_3 \leq \cdots \leq X_i \leq X_{i+1} \leq \cdots \leq X_n. \qquad (3c)$$

In the sense that this sequence is a member of the infinite set (3b), it is just as random as any other. However, I believe that most scientists would never think of it as a random sequence, particularly if n is reasonably large, let us say 10 or more. There are also many orders other than that indicated in (3c) that would likewise not be called random under normal circumstances if observed in the course of *actual experimental work* as contrasted with drawing from a bowl. For example, sequences suggesting functional relationship or marked trends of the variable X with the order would not likely be classed as random; e.g. see fig. 32 (p. 147) and accompanying text.

In other words, given any finite sequence of n terms, it is theoretically possible to write down each of the $n!$ different orders that might be expected to occur with equal frequency in the set (3b). The scientist or engineer looking at these $n!$ orders would distinguish several that, if they had occurred in his everyday experience, he would not call random. Pushed for an explanation, he would likely say that the sequences he would pick out of (3b) and not call random would be those that, if they occurred in the course of his work, he would attribute to some nonrandom instead of random causal process. Pushed a little further, he would say that if in practical experimental work he gets one of these orders that he would choose upon the basis of past experience as being nonrandom, and that if he repeats again and again the same physical operation, the new sequences thus obtained will not often be much like the ones he would expect to get had the original finite sequence been drawn from a bowl. For example, if he were to obtain in practical work a trend such as indicated by (3c), he would be more likely to expect the next m observations to suggest the presence of a trend than if the original sequence that showed a trend had been obtained by the random operation of drawing from a bowl. The importance of being able to get clues from the characteristics of a sequence will be seen later (p. 27).

Now we are in position to make three observations that are of fundamental importance in quality control work. First, it appears hopeless to define random order in a useful way for a *specific* sequence. Instead it appears that *the only operationally verifiable way to define random order is in terms of some chosen random operation.* A random sequence in this sense is then simply a member of an infinite class of sequences obtainable through repetitions of the chosen random operation. Second, what the experimentalist, at least the quality control engineer, implies by saying that a sequence is not random is something that can be checked only by making further experiments. For example, the implication may be that further study of the cause system that produced the original finite sequence called nonrandom by the experimentalist will reveal ways by which this cause system may be modified through the process of eliminating assignable causes; or the implication may be something operationally verifiable about future observed values of X given by this cause system that will be characteristically different from that produced by the random operation of drawing from a bowl.

There is of course, on reflection, nothing mysterious about this situation wherein a sequence is called random if it is known to have been produced by a random operation, but is assumed to be nonrandom if occurring in experience and not known to have been given by a random operation. The fact simply is that such an observed sequence may or may not have been given by a random operation, and past experience has shown that such sequences occurring in practice are more likely to have arisen as a result of a nonrandom than of a random operation. In any case, the implication of the statement that an observed sequence is or is not random can be verified only in the *future* and is not one that can be verified by comparing the order in the observed sequence with any or all of the $n!$ orders of the same set of numbers. Third, the experimentalist usually considers the order in any sequence of observed results to be one of the most helpful clues to the physical interpretation of his results as a basis for future predictions. He is forever on the lookout for orders of special importance. When he can no longer distinguish anything significant in an observed order, he is likely to take it for granted that the observed data have been taken under the same essential conditions.

Randomness not verifiable by comparison with other orders

The operation of statistical control to be described shortly is a successful attempt to extend the usefulness of order in an observed sequence as a clue to the making of valid predictions in operationally verifiable terms beyond the place where the experimentalist fails without the aid of a criterion of control to attribute significance to order. To extend the usefulness of observed order in a sequence in this way is a basic objective of the theory of statistical control of quality and constitutes an extension of the significance attached to order so well established in the history of science.

There is an indefinitely large number of ways in which the order in the original sequence may be expressed in terms of the order of subsamples of the original sequence by using an indefinitely large number of different statistics such as average, standard deviation, and all moment functions, to mention only a portion of those that are possible. There is also an indefinitely large number of ways of breaking up the sequence into subsamples. In other words, we might include all of the results of what is generally termed the mathematical theory of distribution, to which contributions are being added daily, as a basis for characterizing the order in such a sequence. Personally, I like to look upon the theory of distribution as providing an indefinitely large reservoir of criteria by which one may describe the order in a sequence characterizing the physical state of statistical control.

There is no unique description of a state of control. In the last few paragraphs we have briefly indicated the many difficulties confronting anyone searching for a *unique description* of the characteristics that a sequence arising from a state of statistical control must have. Evidence has been given for believing that such a unique description can not be found. Attention has been directed to the significance of the order in which the numbers appear in the sequence as constituting a part of the requirement of any definition of a mathematical state of statistical control; yet here again there is no definite unique test that is necessary and sufficient to define the order that the mathematician refers to as random. What then may we conclude about specifying the state of control in mathematical terms?

It is obvious that we can not hope to specify *the* mathematical state of statistical control in a complete manner. All that we can hope to do is to make some arbitrary choice of criteria and some arbitrary choice of random operation such as drawing from a bowl to be taken as characterizing such a state, being careful that each criterion chosen takes into account the order in the sequence. The definition of random in terms of a physical operation is notoriously without effect on the mathematical operations of statistical theory because so far as these mathematical operations are concerned random is purely and simply an undefined term.[11] The formal and abstract mathematical theory has an independent and sometimes lonely existence of its own. But when an undefined mathematical term such as random is given a definite operational meaning in physical terms, it takes on empirical and practical significance. Every mathematical theorem involving this mathematically undefined concept can then be given the following predictive form: *If you do so and so, then such and such will happen.* Hence the process of making a physical application of the mathematical theory consists in specifying the *human operations* by which physical meaning is given to the mathematically undefined terms. We can then proceed to determine if the resultant predictions of physically observable events suggested by carrying out the associated mathematical operations are valid. *For the empirical verification of the usefulness of mathematical statistics, the validity of the assumptions involved in giving a specific physically operational meaning to the term random is of fundamental importance.* Hence it is that for the successful application of statistical theory great care needs to be given to the method of defining the state of statistical control in terms of the physical operations and the associated mathematical operations based upon the mathematically undefined concept of random.

[11] *Cf.* H. Cramer, *loc. cit.*, p. 5; J. Neyman, "Outline of a theory of statistical estimation based on the classical theory of probability," *Phil. Trans. of the Roy. Soc. of London*, vol. A236, pp. 333–380, 1937; in particular, pp. 338–9.

How to build a model of a state of statistical control. Postulate I. Thus far an attempt has been made to indicate some of the important characteristics of the a priori and definitive concept of a state of statistical control in terms of the characteristics of an infinite set of infinite sequences (3*a*) in one case and an infinite set of finite sequences (3*b*) in another, where each set is generated by a physical operation characterized as random. It has been implied that, in quality control work, whenever the system of chance causes producing variations in an observed sequence (3) of some quality characteristic X is such as to produce a sequence that is a member of the class (3*a*), the chance cause system is to be considered as being in a mathematical state of statistical control or in a state where one can build a mathematical model to represent certain characteristics of that particular state. The two functions desired of such a model are that

(1) It shall serve as a computing device in making predictions.

(2) It shall suggest new physical experiments to be made in trying to attain a state of statistical control.

So far as a model can be constructed upon the basis of knowledge of a finite number n of terms of the sequence (3) to provide valid predictions about the remainder of the sequence (3), such a model will obviously be of great use in many engineering applications. In fact the attainment of such a model is the ideal goal in many instances in establishing economic tolerances in the sense to be discussed in chapter II.

For reasons already considered, no model can ever be theoretically attained that will completely and uniquely characterize the indefinitely expansible concept of a state of statistical control. What is perhaps even more important, on the basis of a finite portion of the sequence (3)—and we can never have more than a finite portion—we can not reasonably hope to construct a model that will represent *exactly* any specific characteristic of a particular state of control even though such a state actually exists. Here the situation is much like that in physical science where we find a model of a molecule; any model is always an incomplete though useful picture of the conceived physical thing called a molecule.

In this section we shall consider the simplest case of building a mathematical model in which the observed sequence (3) is drawn one term at a time with replacement and thorough mixing by some one that is blindfolded or, as we shall say, drawn from a bowl containing an unknown population. We have already chosen to characterize this operation of drawing as random, so we can begin at once to construct our model without first testing for randomness as we shall have to do in the next section when we take up the problem of attaining a state of statistical control for a manufacturing process.

All that we shall attempt to do here is to examine briefly but critically the principles underlying the practical procedure in constructing a model.

Distribution theory is purely formal mathematics

First, let us note just what it is that we are to call the model. An important element is the concept of a universe or more accurately a frequency function

$$p' = \int_{\alpha}^{\beta} f(X, \theta'_1, \theta'_2, \cdots, \theta'_s)\, dX \tag{2a}$$

where the indicated integral from α to β gives the relative frequency p' of occurrence of X for the infinite sequence (3) within the interval α, β, the integral from $-\infty$ to $+\infty$ being unity. The other important element of the model is the formal mathematical theory of distribution that gives the rules for deriving other distribution functions. We call these derived functions frequency distributions of statistics of samples of n drawn from the parent universe, but here again the mathematical operations are, from the viewpoint of meaning, independent of what the results are called. The essential fact is that the theory of distribution is purely formal mathematics.

Now let us consider how we are to go from an infinite sequence such as (3) or, more particularly, from a finite portion thereof, to the universe. As already noted above, the mathematician *postulates* that the observed fraction p of values of X within any arbitrary range $\alpha < X < \beta$ for a sample of size n approaches p' as a limit as $n \to \infty$, as indicated in (5). Likewise, the mathematician picks out certain functions θ_i $(i = 1, 2, \cdots, m)$ of any finite portion consisting of n terms (let us say the first n terms) of the infinite sequence, and postulates that the limits

$$\operatorname*{Lim_s}_{n=\infty} \theta_i = \theta'_i \qquad i = 1, 2, \cdots, m \tag{5a}$$

exist. The same limits are usually postulated for all sequences drawn at random from the same bowl. As is well known, of course, there is no accepted way of proving the physical existence of these limits and for that reason I like to indicate this fact [12] by using the symbol Lim_s instead of Lim. For our present purpose, however, we are interested in the physical operation associated with the limit (5a) by which we go from a θ_i to θ'_i, and the significance thereof from the viewpoint of probable inference.

[12] It should perhaps be noted, however, that formal mathematical concepts of limits, continuity, and the like are introduced in much the same way in many physical theories where our concept of the physical condition described does not rigorously satisfy the conditions implied by the concepts. For example, continuity is a fiction so far as its use in physical theory is concerned, yet the use of such a concept is often justified by the results obtained.

Drawings from a bowl; does $\bar{X}$ approach a limit, statistically? There is no answer

Physically, perhaps the nearest that one can approach the nature of a statistical limit is by drawing with replacement from an experimental universe written on a series of "physically similar" chips,[13] the ideal bowl experiment previously mentioned (p. 9). Fig. 5 shows one observed approach. The distribution of the numbers written on the chips in the bowl was approximately normal [14] and symmetrical about zero as an arithmetic mean. The ordinate of each point is the observed average $\bar{X}$ for the sample of size n corresponding to the abscissa of that point; as n varies,

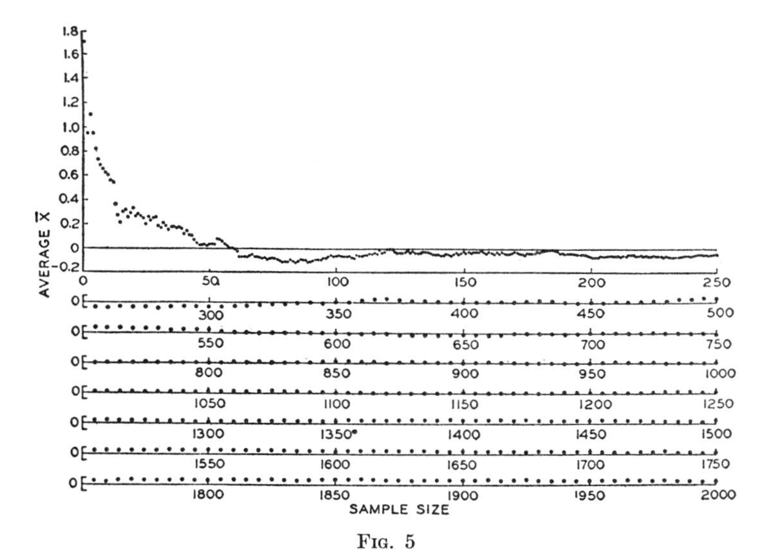

Fig. 5

the average $\bar{X}$ varies. It is of interest to note how the observed average swings back and forth about zero, which is sometimes spoken of as the theoretical limit. Do we know that with increasing sample size the average $\bar{X}$ in this particular case approaches any particular value $\bar{X}'$ in the sense of a statistical limit? No matter how many observations we might take, I should still not know how we could answer this question with certainty.[15]

[13] See the reference to one such bowl experiment cited in a footnote on page 10.

[14] Any actual distribution must of course be discrete and have definite cut-offs in the tails. An exactly normal distribution is unrealizable, but is a mathematical artifice to facilitate calculations. In this problem the actual distribution of the numbers on the chips is of no consequence, for we are concerned only whether $\bar{X}$ approaches a limit statistically, the actual value of the limit, if it exists, being of no importance at present. *Editor.*

[15] On this point see further discussion in chapter IV.

One might ask whether this approach satisfies that symbolized formally by equation (5*a*). I know of no way of answering this question in an operationally definite way any more than I know of a way of checking once and for all a sequence in an operationally definite way to see if it represents a statistical state of control. If we assume that the dotted curve in fig. 5 approaches a limit, the practical significance of this conclusion is that we are tacitly adopting the empirical rule of inference that an average of n observations is to be taken in preference to an average of $n - 1$.

Likewise, in the practical operation of setting up the model, we assume that we can approach the functional form of f and the parameters therein by acting as though the limits (5) and (5*a*) exist, or more accurately, by the rule of taking a p or a θ_i calculated from $n + 1$ terms of an infinite sequence such as (3) in preference to a p or θ_i calculated from n such terms. Stated more generally, this amounts to basing action on the following fundamental

> ***Postulate I.*** **A model of a statistical state based upon $n + 1$ terms of a sequence defined as random is to be chosen instead of a corresponding model based upon n terms.**

Of course, it should be kept in mind that the process of setting up a model of a state of control in the way just described is up to this point [16] limited to the case where the sequence (3) is drawn from a bowl and hence is given by what we have chosen to define as a random operation. Such an operation characterizes a physical state of statistical control representing the limit to which one may hope to go in attaining valid predictability and a state where the one making the drawings as prescribed can not do anything to control the limits of observed variability. It must, however, be kept in mind that logically there is no *necessary* connection between such a physical statistical state and the indefinitely expansible concept of a statistical state in terms of mathematical distribution theory. There is, of course, abundant evidence of close similarity *if* we do not question too critically [17] what we mean by close. What is still more important in our present discussion is that if this similarity did not exist in general, and if we were forced to choose between the formal mathematical description and the physical description, I think we should need to look for a new mathematical description instead of for a new physical description because the latter is apparently what we have to live with. It is the practical man's good fortune that mathematical distribution theory seems to agree so closely with what he gets in drawings from an ideal experimental universe. As an indirect result, distribution

[16] In the next section we shall see how through the operation of control, sequences can be attained that may also be treated as random from the viewpoint of constructing a model.

[17] See J. Neyman, *Lectures and Conferences on Mathematical Statistics* (The Graduate School, The Department of Agriculture, Washington, 1938), pp. 19–32.

theory (mathematical statistics) must become the stock in trade of the control engineer.

STATISTICAL CONTROL AS AN OPERATION [18]

Let us first see what the operation of control is designed to do. The statistician looking at the function or purpose of the operation of control will likely see it as a procedure for attaining a state of statistical control of some variable whereas the engineer will see it as a means of effecting certain economies and attaining the highest degree of quality assurance at a given cost. Presumably both the statistician and the engineer are interested in understanding the operation of control as a scientific procedure. In what follows, an attempt is made to present the important characteristics of the operation from each of these viewpoints.

The aimed-at value C; the action limits A and B

In the beginning of this chapter we noted the steps that had been taken in going from the concept of an exact fit of interchangeable parts based upon the concept of an exact science, to the concept of tolerances, fig. 4, p. 6. Statistical theory then stepped in (1924) with the concept of two action or control limits A and B that lie, in general, within L_1 and L_2, as shown in fig. 6 (next page). These limits are to be set so that when the observed quality of a piece of product falls outside of them, even though the observation be still within the limits L_1 and L_2, *it is desirable to look at the manufacturing process in order to discover and remove, if possible, one or more causes of variation that need not be left to chance.* In other words, whereas the limits L_1 and L_2 provide a means of gauging the product *already made,* the action limits A and B provide a *means of directing action toward the process* with a view to the elimination of assignable causes of variation so that the quality of the product *not yet made* may be less variable on the average.

Furthermore, the statistical theory of quality control introduces the concept of the expected value C lying somewhere between the action limits A and B. This point C serves in a certain sense as an aimed-at value of quality in an economically controlled state. We might pause a moment to note the importance of the point C from the viewpoint of design or the use of material that has already been made. Let us take, for example, a very simple problem of setting overall tolerance limits. Suppose that we start with the concept of the go, no-go tolerances of 1870 (fig. 4, p. 6) and that

[18] This subject is discussed at length in my book, *Economic Control of Quality of Manufactured Product* (Van Nostrand, New York, 1931). It is also discussed in a most helpful way in *The Application of Statistical Methods to Industrial Standardization and Quality Control* by E. S. Pearson (British Standards Institution, London, 1935) and in the *Manual on Presentation of Data* (American Society for Testing Materials, 260 S. Broad St., Philadelphia, 1933).

we wish to fix the overall tolerance limits for n pieceparts assembled in such a way that the resultant quality of the n parts is the arithmetic sum of the qualities of the component parts. An extremely simple example would be the establishment of tolerance limits on the thickness of a pile of n washers or, in general, any n laminated pieceparts in terms of the tolerance limits on one. The older method of fixing such limits was to take the sum of the tolerance limits on the individual pieceparts, but the tolerance range resulting from such practice is usually many times larger than it needs to be. The economical way of setting such tolerance limits for a product in a state of statistical control is in terms of the concept of the expected value C of the quality, and the expected standard deviation about this value. The concept of the expected value is of fundamental importance in all design work in which an attempt is made to fix overall tolerances in terms of those of pieceparts.

Thus we see that for reasons of economy and quality assurance it is necessary to go beyond the simple concept of the go, no-go tolerance limits of the customary specification and to include two action limits A and B and an expected value C, as shown schematically in fig. 6. Statistical theory alone is responsible for the introduction of the concept of the action limits A and B and the expected value C.

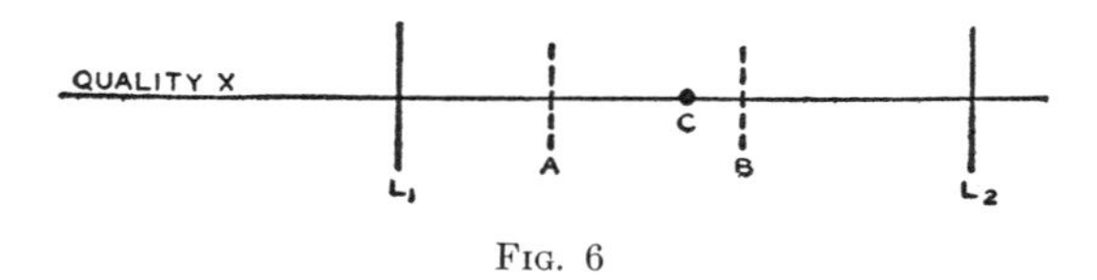

FIG. 6

It should be noted that if there were no reason connected with economy or quality assurance for going beyond the concept of the go, no-go tolerance limits, statistical theory would have nothing to add. Likewise, it should be noted that, although the action limits A and B may lie within the tolerance limits L_1 and L_2, the product already produced and found by inspection to be within the limits L_1 and L_2 is still considered to conform, even if outside A and B. In other words, the action limits A and B do not apply as a gauge for product already made: their function is to call attention to evidence for believing that the manufacturing process includes assignable causes of variation in the quality that may give trouble in the future if they are not found and removed.

The operation of statistical control. The use of statistical techniques in the way just described introduces a modification in the customary operation of control and in this sense constitutes an "operation of statistical control" directed toward the attainment of a state of statistical control.

The specification of an operation of statistical control consists of the following steps:

> **1. Specify in a general way how an observed sequence of n data is to be examined for clues as to the existence of assignable causes of variability.**
>
> **2. Specify how the original data are to be taken and how they are to be broken up into subsamples upon the basis of human judgments about whether the conditions under which the data were taken were essentially the same or not.**
>
> **3. Specify the criterion of control that is to be used, indicating what statistics are to be computed for each subsample and how these are to be used in computing action or control limits for each statistic for which the control criterion is to be constructed.**
>
> **4. Specify the action that is to be taken when an observed statistic falls outside its control limits.**
>
> **5. Specify the quantity of data that must be available and found to satisfy the criterion of control before the engineer is to act as though he had attained a state of statistical control.**

In the next few paragraphs I shall consider briefly each of these steps and indicate the nature of the available evidence to show that the operation as a whole successfully accomplishes its objective in practice.

Let us think of a particular manufacturing process as an operation of making a given kind of object, and let us assume as above that this operation can be repeated again and again at will. Let us assume that we want to attain a state of statistical control of some quality characteristic X; that n pieces of the product have been made; and that the qualities of these n pieces in respect to the characteristic X are available in the order that the pieces were produced. These n values of X may be thought of as constituting the first n terms of an infinite sequence (3) corresponding to what we should get under similar conditions by repeating again and again the operation of production.

It is essential for an understanding of the operation of control that we distinguish three kinds of acts that are involved. These are (*a*) mental operations or judgments typical of which is the judgment that two or more observations are made under the same or different conditions, (*b*) mathematical operations such as are involved in constructing a criterion of control, and (*c*) physical operations such as looking for an assignable cause when an observed point fails to satisfy a criterion of control.

Some comments on the first step in the operation of control. The importance of order. In order to take this step, we must decide how the original set of n data is to be used as a clue to the existence of assignable causes of variability. We start with the assumption that when the operation of production is random, that is, when it is in a state of statistical control, there are no assignable causes present in the production process. Hence our clue to the existence of assignable causes is anything that indicates nonrandomness. However, as already pointed out above, any set of n values of X considered as a sample or as a sequence might have been obtained by some random operation. Likewise it might have been obtained by a nonrandom operation. We must therefore take into account the previously observed fact that there is no unique test for randomness of the cause system producing the data in terms of the n observed data.

If a set of n data is to serve as a clue to the state of control, two conclusions are obvious. First, we must depend on past experience to suggest what, if any, characteristics of a given set of n values of X are more likely to occur in nature as a result of a nonrandom than as a result of a random operation. For example, if you were told that nine successive determinations of the density of oxygen gave in proper units 1.42891, 1.42892, 1.42892, 1.42894, 1.42894, 1.42895, 1.42895, 1.42896, 1.42900, you would likely suspect a nonrandom condition. It should be kept in mind that from the viewpoint of the operation of control, nonrandom is a category for a temporary pigeon-hole for those states of control that we are to look at further in an attempt to find assignable causes of variability. If, however, you were given the following order in which they actually did occur,[19] 1.42900, 1.42894, 1.42896, 1.42892, 1.42895, 1.42891, 1.42892, 1.42894, 1.42895, you would likely conclude that the sequence represented a random condition. Second, the acceptance of any specific characteristic such as order in any given set of n observed values of X as an indication of the state of control can only be confirmed in the light of *future* experience. Hence the only operationally verifiable way in which the set of n observed data may serve as a clue to the state of control is by serving as a link between past and future experience.

Keeping in mind that our principal object is to detect the presence of assignable causes of variability, it is natural to try to make use of the fact that there are certain identifiable orders observed in experience that are not equally likely to be associated in future experience with every other possible order. This is the more reasonable since every scientist and engineer well knows that some observed orders are much more likely to be produced by nonrandom than by random operations. In fact it is only when he has difficulty in distinguishing orders of this kind—trends, cyclic movements, functional relationships, and erratic effects—that he appeals to the statistician.

[19] See the *Journal of the American Chemical Society*, vol. 61, pp. 223–228, 1939.

There are, however, other reasons for examining the observed order for clues of nonrandomness. For example, the identifiable order of three or more numbers, whence comes our concept of the ordinal numbers, is based on the property of "betweenness" and not on the absolute values of the numbers; so likewise the significance of observed order is independent of the frequency distribution of the observed set of numbers as well as that of the potentially infinite sequence of which the n observed numbers constitute but a finite part.

Of course, if we neglect the significance of observed order, we may still ask whether the set of n observed values is a likely random sample from some assumed universe; and if it is not a likely sample in this sense, we may reject the hypothesis that it came from such a universe. However, in applying such a test we must first introduce an assumption that the observed order is given by a random operation so that we may arrive at the functional form of the universe and at estimates of the parameters. Abundant evidence will be presented in the following chapters to show that such an assumption is almost never justified in practice; hence we are confronted with the necessity of assuring ourselves that we are dealing with a sequence given by a random operation before we can justify the customary interpretation of a test of this character. Under such conditions, it appears that certain requirements on the order of happening are *primitive*. Moreover, such a test to determine whether an observed set of n data is a likely random sample from some assumed universe does not in itself indicate whether the observed sample is likely to have arisen from some nonrandom operation. Finally, such a test neglects the significance of the observed order as a clue to nonrandomness.

Some comments on the test of a hypothetical universe

Considerations of the character indicated in the last few paragraphs have indicated the need, in quality control work, of stressing the significance of certain characteristics of the observed order as clues to nonrandom operations in the production process. The fact that the successful choice of the observed orders most likely to indicate nonrandomness (or the presence of assignable causes) must be based on experience simply emphasizes the importance of a broad experience in the subject matter of any given field as a background for establishing efficient tests for assignable causes.

Some comments on the second step in the operation of control. Let us start by rewriting the infinite sequence (3) in the form

$$\begin{array}{cccccccccccc} X_1, & X_2, & \cdots, & X_i, & \cdots, & X_n, & X_{n+1}, & \cdots, & X_{n+j}, & \cdots \\ | & | & & | & & | & | & & | & \\ C_1 & C_2 & & C_i & & C_n & C_{n+1} & & C_{n+j} & \end{array} \qquad (3d)$$

where the symbol C_i ($i = 1, 2, \cdots$) stands for the condition under which the associated X_i was obtained and where the term condition is used as it is in the phrase "same essential conditions." We may symbolize the judgment that the conditions are essentially the same by the expression

$$C_i \approx C_j. \tag{6}$$

In the initial stages of the operation of any production process, it is likely that the engineer will not be willing to assume that $C_i \approx C_j$. Instead, he will likely point out what he considers may be important differences in the conditions, as for example, differences in source of raw materials, variations in humidity, wear of tools, and the like. True enough, upon extended study some or all of these differences in conditions may not be found to be assignable causes of variability in the X's. However, to begin with, these differences in conditions constitute our best clues to what on further study may prove to be assignable causes. Hence it is desirable that the engineer or scientist provide a means of grouping the X's upon the basis of observed differences in conditions that may later prove to be assignable causes. It is important to note that from the viewpoint of interpretation, the grouping of the X's in this manner is independent of their magnitudes.

Every scientist and engineer follows such a procedure as a part of his daily work; if the measurements broken up into subgroups in this way are radically different from one group to another, the conclusion is usually drawn that the corresponding differences in conditions constitute assignable causes of variation in the quality characteristic X, and this conclusion is reached without ever calling on the statistician for advice. If, on the other hand, the subgroups of X's are not "clearly" different and show some overlapping, two alternative courses are open to the scientist or engineer. One of these is to conclude that the conditions are essentially the same and the other is to call in the statistician to advise whether the observed differences between the groups of X's are likely to have arisen as sampling fluctuations under a state of statistical control. Thus we come upon the problem, long familiar to the statistician, of trying to devise a statistical test for determining whether two or more samples are significantly different. The statistician may feel that he is now on familiar ground and make use of some statistical test for significance. However, in doing so, he usually introduces the assumption that the values of X in any particular subgroup proposed by the experimentalist constitute a random sample from *some* universe, and he usually assumes that the functional form of this universe is normal. Of course, the mathematical statistician can then make certain statements that follow rigorously from the assumptions. Obviously, however, the practical importance of any deductions of this character depends upon whether the assumptions are representative of the actual conditions.

To illustrate, let us consider the significance of the assumption that each subgroup constitutes a random sample from some universe. Obviously *any* finite set of numbers *might* be given by a random operation, and would constitute a random sample from some universe. In fact, any finite set of numbers might be a sample from a normal universe. Statistical theory enables us to say rigorously some things of interest about how certian characteristics of successive samples from the same normal universe may be expected to vary even though we do not know the average $\bar{X}'$ and standard deviation σ' of the universe.

At this point one of the distinguishing characteristics of the control statistician shows up. Irrespective of the result of applying any statistical test for significant differences between subgroups selected by the experimentalist solely upon the basis of his knowledge of the conditions under which the values of X were obtained, the control statistician knows that past experience does not justify him in believing that any such subgroup is a random sample of the production process if the only evidence for this belief is that the experimentalist considers the observations in each subgroup to have been taken under the same essential conditions. That is to say, experience has shown that the judgment represented by expression (6), that the conditions underlying a set of n values of quality are essentially the same, is not by itself a satisfactory criterion of randomness. If we could rely on such a judgment as a sufficient condition for believing that a state of statistical control had been reached, there would be no story to tell about the operation of statistical control with which we are concerned in this section.

On the other hand, no scientist or engineer would think for a moment of ignoring the importance of the human judgment that the conditions under which a set of n measurements were made did or did not remain the same. In the successful development of any operation of statistical control, we can not do without the human judgments about the conditions C', but we can not get along solely with them, either. We must seek in addition some criterion of control that makes use of the numerical magnitudes of the observed qualities.

Some comments on the third and fourth steps in the operation of control. Practical requirements imposed on the criterion of control. Criterion I. We shall consider these two steps together because they are so closely interrelated. In fact, what is to be done in step three depends to a large extent upon the action that is to be taken in step four. For example, step four consists of looking for an assignable cause of variability whenever the observed statistic chosen in step three falls outside its control limits. Hence the criterion of control should be as nearly as possible such that when and only when a statistic falls outside its control limits it will be possible to find an assignable cause of variation. And it is to be remembered that if an assignable cause is found and removed, a change (narrowing) of the control limits is required (note paragraph iii on p. 35). Thus the third and fourth steps are closely dependent on each other.

We are now in a position to set down some of the important practical requirements imposed upon the criterion of control, and this we shall do before passing on to comments regarding the fifth step in control.

1. Our criterion of control should indicate the presence of assignable causes of variation.
2. It should not only indicate the presence of assignable causes but also should do this in a way to facilitate the discovery of these causes.
3. It should be as simple as possible and adaptable in a continuing and self-corrective operation of control.
4. It should be such that the chance of looking for assignable causes when they are not present does not exceed some prescribed value.

I have discussed elsewhere what has been termed Criterion I of control.[20] There is no intention of repeating here what was said at that time but in the next few paragraphs an attempt will be made to explain in more detail some of the reasons why this criterion was chosen. Let us see how it meets the four practical requirements just noted.

(i) The principal function of the chart is to detect the presence of assignable causes (1st requirement). Let us try to get clear on just what this means from a practical and experimental viewpoint. We shall start with the phrase "assignable causes." An assignable cause of variation as this term is used in quality control work is one that can be found by experiment without costing more than it is worth to find it. As thus defined, an assignable cause today might not be one tomorrow, because of a change in the economic factors of cost and value of finding the cause. Likewise, a criterion that would indicate an assignable cause when used for one production process is not necessarily a satisfactory criterion for some other process. Obviously there is no a priori, formal, and mathematical method of setting up a criterion that will indicate an assignable cause in any given case. Instead, the only way one can justify the use of any criterion is through extensive experience. The fact that the use of a given criterion must be justified on empirical grounds is emphasized here in order to avoid the confusion of such a criterion with a test of statistical significance. We shall return to this point in some of the paragraphs below. Here it must suffice to recall that any test of statistical significance is a *deductive* inference upon

[20] W. A. Shewhart, *Economic Control of Quality of Manufactured Product* (Van Nostrand, New York, 1931), ch. XX.

the basis of certain fundamental assumptions, and theoretically can be made with any desired degree of exactness. In general, any such test consists in defining some statistic θ of a random sample of n from some assumed universe and computing the probability of getting an observed value of θ outside any chosen range $\theta_1 \leq \theta \leq \theta_2$. Then some arbitrary choice of probability is made and the associated values of θ_1 and θ_2 are computed. An observed value of θ is then defined as significant if it falls outside the corresponding range $\theta_1 \leq \theta \leq \theta_2$. Such a process is deductive. In contrast, when an observed statistic falls outside its control limits, the inductive inference is implied that an assignable cause is present. To check this inductive inference, we must appeal to empirical evidence.

The next point to note is that in developing a control criterion we should make the most efficient use of the order of occurrence as a clue to the presence of assignable causes. The importance of order as such a clue has already been considered (pp. 25–27 ff). As an example, let us consider a case where we have a sequence of n numbers taken under presumably the same conditions. One such set of 204 observations of insulation resistance [21] may be used here to illustrate some of the characteristics of a control chart criterion as a tool for detecting the presence of assignable causes. Grouping these 204 observations into subgroups of four taken in the order in which the observations were made, and applying the control chart Criterion I to the 51 subgroup averages, we get the results shown in the upper half of fig. 7. Here we see indications of the presence of assignable causes of variability, which further research revealed and removed (see pp. 114–5).

Now let us see what would have happened if we had not known the order in which the 204 pieces of insulating material were made. For example, suppose that these pieces had been thoroughly mixed together in a box or tray before the measurements of resistance had been made, as is a very common practice. The 204 measurements of resistance on the 204 pieces of material after they had been thoroughly mixed would have been the same, but we should then know nothing about the order in which the pieces were made. Of the 204! different orders that might be obtained by such a random operation, the order of manufacture, which is the basis of the control chart in the upper half of fig. 7, is no more unlikely than any other. Instead of mixing the pieces of insulating material in a tray or box, and measuring one piece at a time upon drawing it, we may write the 204 original measurements on as many physically similar chips, mix the chips in a bowl, and draw them one at a time without replacement. Suppose we apply the same Criterion I to one sequence of 204 numbers obtained in this way. The results are shown in the lower half of fig. 7. There is no indication of the presence of assign-

More on the importance of order

[21] The 204 observations constitute table 7, given in chap. III, p. 90.

able causes. If in this case the original order had not been given and we had taken instead the order actually given by the random operation of drawing the 204 numbers one at a time from a bowl, the application of

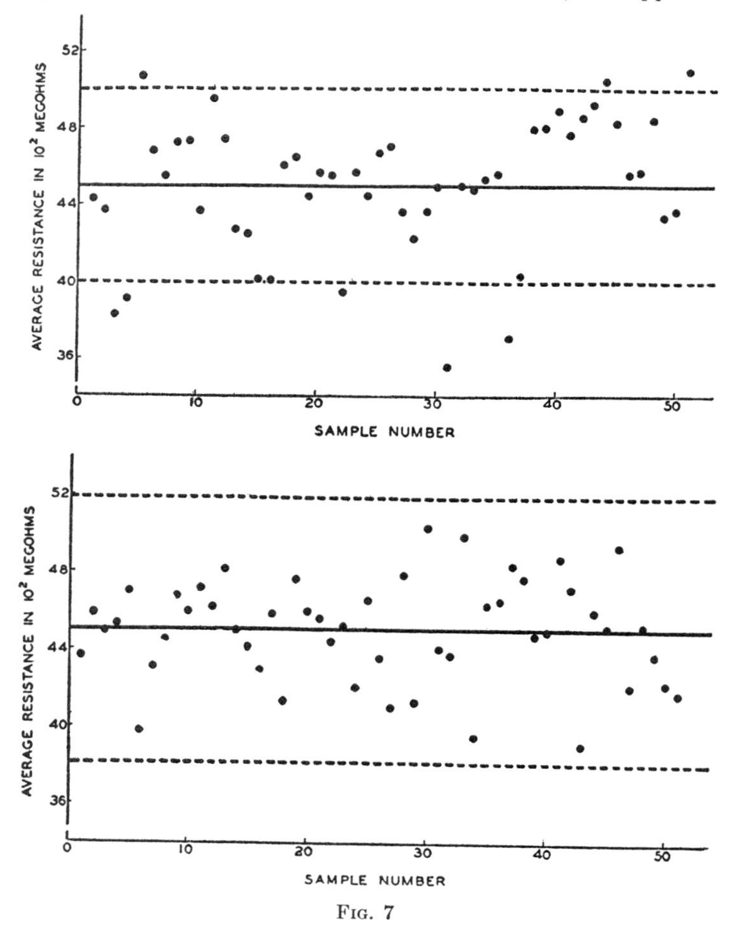

Fig. 7

Criterion I would have failed to detect the presence of assignable causes. And it may be shown theoretically that if we were to apply Criterion I in the same way to all of the 204! possible different sequences, most of them would give no indication of the presence of assignable causes in the sense of showing averages of four outside of control limits. On the other hand, we must remember that the original sequence is one of the 204! possible sequences generated by such a random operation. Hence the failure to meet the criterion does not serve to pick any one of the 204! sequences drawn from a bowl as being nonrandom, because in fact they all are obtained by means of a random operation.

Why then may we place faith in Criterion I as a good indicator of assignable causes, or of those that can be found? As already suggested earlier in this chapter in the discussion of the meaning of random, extensive experience has shown that one almost never finds in practical work an observed sequence even when obtained under presumably the same essential conditions that will satisfy Criterion I, and if assignable causes are looked for when an observed statistic goes outside its control limits such causes are almost always found. If the process of finding and removing assignable causes is continued, we gradually approach a condition where an observed statistic only seldom goes outside of its limits, and if one looks for assignable causes in these rare instances, such causes are not usually found.

It is important to note that in the use of Criterion I to detect the presence of assignable causes, emphasis always has been and must be laid upon breaking up the original sequence into subgroups of comparatively small size. If this is not done, the presence of assignable causes will very often be overlooked. Incidentally the necessity of using small subgroups is not imposed by the particular choice of the criterion used. It would be equally necessary, for example, if we were to use the analysis of variance test instead of Criterion I. Thus if the 51 samples of 4 are analyzed by the analysis of variance, using a probability level of .01 as a test for assignable causes, we get indications of the presence of such causes. If instead of 51 subgroups of 4, we take 4 subgroups of 51, both Criterion I and the analysis of variance test happen to give positive indications in this particular instance, although this is exceptional for subgroups of this size. If, however, we go to 2 subgroups of 102 each, both tests fail. Needless to say, this one example is introduced not to prove the importance of using small subgroups in the criterion of control but simply to illustrate what is usually found in practice.

Small subgroups required in Criterion I

It is reasonable to expect that one may detect more readily the presence of the customary kinds of assignable causes by breaking up the total number of available observations into small subgroups than by breaking them up into larger subgroups. One of the principal reasons is that assignable causes are for the most part those that come and go in an erratic fashion. For example, let us think of an infinite sequence:

$$\begin{array}{cccccccccccccc} X_1, & X_2, & \cdots, & X_i, & X_{i+1}, & \cdots, & X_{i+j}, & \cdots, & X_n, & X_{n+1}, & \cdots, & X_{n+k}, & \cdots \\ | & | & & | & | & & | & & | & | & & | & \\ C_1, & C_2, & & C_i, & C_{i+1}, & & C_{i+j}, & & C_n, & C_{n+1}, & & C_{n+k}, & \end{array} \quad (3d)$$

An assignable cause may, for example, come into the condition C_i and remain present in the next $j - 1$ conditions. The same assignable cause may come in again and again at other places in the sequence. Generally there are several assignable causes of this character present in any production process or physical experiment even when it is judged that the conditions are being maintained essentially the same. By using large subgroups we tend to get overlappings of the effects of different assignable causes, and the effects of a single cause are thus masked. Experience and theory both indicate that a subsample size of four is effective in the majority of instances that have come to my attention. Enough has been said to indicate that an important factor in choosing a criterion of control to indicate the presence of assignable causes is the method of feeding the data into the criterion.

Importance of the method of feeding the data into the criterion. Criterion I uncovers not only assignable causes, but also trends and periodic fluctuations

In practice, Criterion I is useful in detecting the presence of assignable causes not only when a statistic falls outside its control limits, but also when the graphical record suggests the presence of either a trend or a periodic effect, even though the observed values of the statistics for the available subsamples do not fall outside the control limits. For example, a sequence of averages of subsamples of four sometimes reveals such effects as would not be suggested by the original sequence. However, it is obviously not feasible to give any definite rule for the use of such apparent trends and periodic fluctuations with the same assurance that one applies the rule of looking for an assignable cause whenever an observed statistic in a subsample falls outside its control limits.

(ii) Next let us consider the second requirement of a criterion of control, namely, that it shall not only indicate the presence of an assignable cause but that it shall do this in a way to facilitate the discovery of the cause. Obviously if an assignable cause is indicated, we must be able to put our finger on the conditions existing at the time the cause is present if we are to find the cause and remove it. Again, with regard to the infinite sequence (3*d*), if an assignable cause is present in the conditions C_i to C_{i+j}, the control criterion should indicate its presence in this set of conditions. Criterion I in the form of a control chart is designed to meet this requirement.

We are now in a position to see another practical advantage of using small subgroups. Assume for the moment that in analyzing the set of 204 values of insulation resistance we make use of 4 subgroups of 51 each instead of the 51 subgroups of 4 each shown in fig. 7. Obviously it is much more indefinite to know only that an assignable cause entered during the time that a subgroup of 51 pieces of material was being made than to know that it entered during the time that a subgroup of 4 pieces was being made.

Another reason for small subgroups

(iii) We next come to the third-mentioned requirement, namely, that the criterion shall be as simple as possible and adaptable to a continuing and self-corrective operation of control. Experience shows that the process of detecting and eliminating assignable causes of variability so as to attain a state of statistical control is a long one. From time to time the control chart limits must be revised as assignable causes are found and eliminated. The continuing control chart record showing a succession of modifications presents a complete and up-to-date history of the available evidence for indicating the progress that has been made up to the present in the process of attaining control.

A simple procedure is used for establishing the limits without the use of probability tables, because it does not seem that much is to be gained

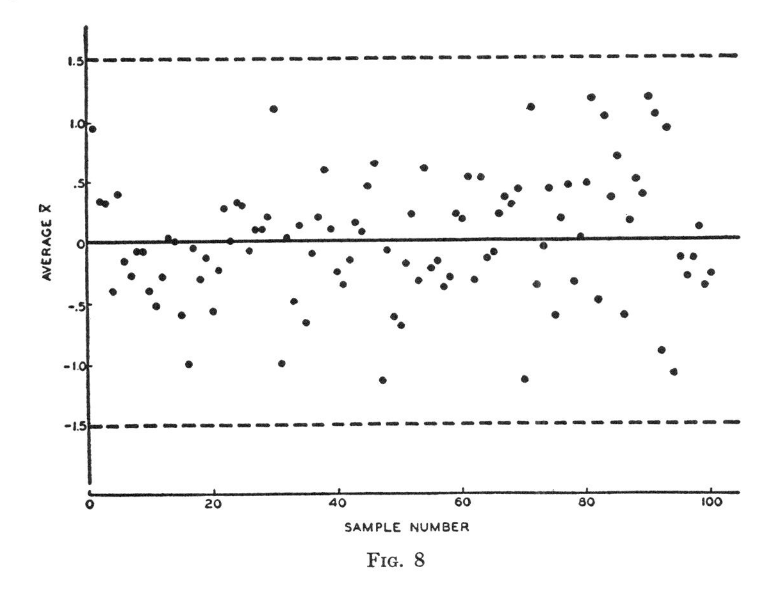

FIG. 8

during the process of weeding out assignable causes in order to attain a state of statistical control by trying to set up exact probability limits upon the basis of assumptions that we know from experience do not hold until the state of statistical control has been reached. This is particularly true since such probabilities do not indicate the probability of detecting assignable causes but simply the probability of looking for such causes when they do not exist, which is of secondary importance until a state of statistical control has been reached. Then too, as already indicated, the design of an efficient criterion for the important job of indicating the presence of assignable causes depends more upon the method of breaking the sequence up into subgroups of a given size taken in a certain order than it does upon the use of any exact mathematical distribution theory.

(iv) True enough, as we approach closer and closer to a state of statistical control it becomes important to have a criterion that does not indicate trouble too often when such trouble is not present. This is the fourth requirement as listed above. The control limits as most often used in my own work have been set so that after a state of statistical control has been reached, one will look for assignable causes when they are not present not more than approximately three times in 1000 subsamples, when the distribution of the statistic used in the criterion is normal. For example, fig. 8 shows Criterion I applied to a sequence of 100 averages of four corresponding to a sequence of 400 drawings with replacement from a normal universe. Not one of the 100 averages falls outside the limits although in the long run we should expect about three in 1000 to fall outside.

Even in trying to keep the probability of looking for assignable causes when they are not present below some limiting value, it is necessary to make some considered choice depending largely upon the costliness of thus looking unnecessarily for trouble. Since there is no a priori exact basis for making this choice it is felt that the simple rules of setting control limits as described in the literature are satisfactory.

Some comments on the fifth step in the operation of control. Even after we have found a suitable criterion of control there remains an exceedingly important practical question to be answered: how long a run of observations satisfying the criterion of control must we have before we can rest assured for practical purposes that a state of statistical control has been attained? Suppose we applied such a criterion to a short sequence of observed values, let us say a sequence of eight, and got no evidence of the presence of assignable causes; should we conclude solely upon this evidence that the process or operation giving rise to the observed values is in a state of statistical control? The answer given by experience in quality control work is definitely *No*. For example, I have never found an instance where, if it had been concluded that a state of statistical control had been reached solely on the basis of evidence provided by a small sample, such a conclusion would not later have been shown to be false. Thus if we apply Criterion I to the first two samples of four of the sequence of 204 observed values of resistance discussed above, we get no indication of the presence of assignable causes. Nevertheless the *process* of making the pieces of insulating material was not in a state of statistical control, as later work revealed, although it may have been that no assignable cause was present *during the time these first eight pieces were being made.*

Before going further, we should note the fundamental and very important difference between an inference that a criterion of control when applied to a sequence of data does not indicate the presence of assignable causes, and an inference that a state of statistical control has been reached upon the evidence that a criterion of control when applied to a given finite sequence does not indicate the presence of assignable causes. As previously noted (p. 33) it is found that assignable causes may again and again come into and go out of a production process or any physical operation repeated an indefinitely large number of times under presumably the same essential conditions. It is therefore possible that no assignable cause is present during the time that a finite sequence is being taken, but this in itself does not necessarily mean that a state of statistical control has been reached or, in other words, that all assignable causes have been eliminated from the process considered as an operation that can be repeated at will an indefinitely large number of times. My own experience has been that in the early stages of any attempt at control of a quality characteristic, assignable causes are always present even though the production operation has been repeated under presumably the same essential conditions. As these assignable causes are found and eliminated, the variation in quality gradually approaches a state of statistical control as indicated by the statistics of successive samples falling within their control limits, except in rare instances, and by the fact that when assignable causes are looked for in these rare instances they are seldom discovered. It has also been observed that a person would seldom if ever be justified in concluding that a state of statistical control of a given repetitive operation or production process had been reached until he had obtained, under presumably the same essential conditions, a sequence of *not less than twenty-five samples of four that satisfied Criterion I.* In certain instances, where it is for some economic or other kind of reason essential that we be practically certain that we have attained a state of statistical control, it may be desirable to have a longer sequence of samples of four. For example, if one wants to attain economic minimum tolerances for a given quality characteristic based upon the assumption that the production process is in a state of statistical control, it may be necessary, as we shall see in the next chapter, that a total sample size of not less than one thousand give no indication of the presence of assignable causes.

Assignable causes of variation are almost always present in the early stages. They may come and go, and the attainment of statistical control is a gradual process. A long sequence is required

The operation of statistical control as a whole. We are now in a position to view at better advantage the operation of statistical control as a whole. As has already been noted, this operation is a *continuing, self-corrective one* designed for the purpose of attaining a state of statistical control. The operation itself must not be confused with the criterion of control: the operation of control not only indicates *how* the data are to be broken up and fed into the criterion of control and what action is to be taken when an observed statistic falls outside its control limits, but also indicates *how many* data must be fed into the control criterion without getting any evidence of assignable causes before the control engineer is to *act* as though he had attained a state of statistical control. The operation of control is in this sense a dynamic process involving a chain of actions, whereas the criterion of control is simply a tool used in this dynamic process. The successful quality control engineer, like the successful research worker, is not a pure reason machine but instead is a biological unit reacting to and acting upon an ever changing environment.

The operation of statistical control is not to be confused with the criterion of control

Example of what can be done in practice. It may be helpful to look at a typical example illustrating how the operation of control works in practice. Fig. 9 shows a control chart for averages of 136 successive samples. The

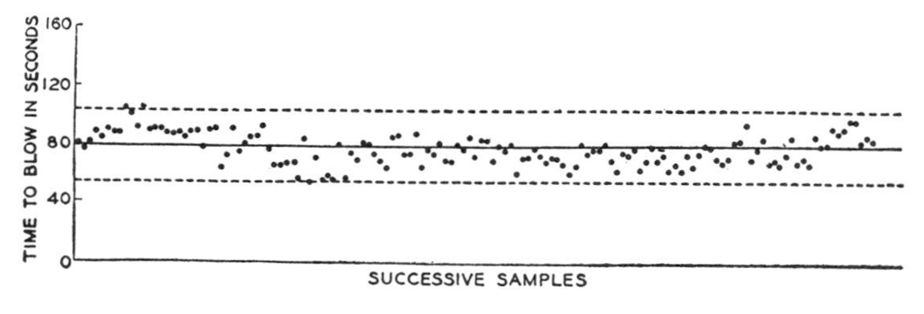

Fig. 9

quality characteristic is the blowing time of a certain kind of fuse. In the preliminary survey which took place prior to the taking of these data, assignable causes were indicated and removed, and the manufacturing process brought into a state of control. This chart is a typical illustration of the fact that once we attain a condition of control, in which a comparatively long sequence of averages of small subsamples taken under presumably the same conditions remains within the limits of Criterion I, this condition usually continues. The averages of the subsamples remain within the control limits almost as well as if the samples had been obtained from a normal bowl universe! That such a state of control can be attained under commercial conditions is all the more impressive when in the next chapter we find that some of the most precise measurements of physical science do not meet this stringent control chart test.

Two kinds of errors in the operation of control. Since a scientific inference about experience can never be more than probable, it is always subject to two general kinds of errors which we may write as follows:

e_1 Sometimes when a scientific hypothesis H is rejected, the hypothesis H is nevertheless true.

e_2 Sometimes when a scientific hypothesis H is accepted, the hypothesis H is nevertheless false.

Neyman and Pearson have considered specific instances of these two general kinds in testing certain statistical hypotheses.[22] They consider the problem of having been given a sample consisting of the first n terms of an infinite sequence considered without respect to order, to determine whether it came from a universe π (hypothesis A). Representing the set of n values as a point Σ in hyperspace, they say—

> *Setting aside the possibility that the sampling has not been random* or that the population has changed during its course, Σ must either have been drawn randomly from π or from π', where the latter is some other population which may have any one of an infinite variety of forms differing only slightly or very greatly from π. The nature of the problem is such that it is impossible to find criteria which will distinguish exactly between these alternatives, and whatever method we adopt two sources of error must arise:
>
> e_{11} Sometimes when Hypothesis A is rejected, Σ will in fact have been drawn from π.
>
> e_{21} More often, in accepting Hypothesis A, Σ will have been drawn from π'.

These two kinds of errors are called by Neyman and Pearson "errors of the first and second kinds," and are obviously somewhat like two different pairs of errors encountered in using the operation of statistical control.

The first of the two pairs of errors (e_1 and e_2) is encountered in interpreting a criterion of control applied to a given finite sequence of observations, and may be written in the following form—

e_{12} We may reject the hypothesis that there existed, at the time the finite sequence was obtained, one or more assignable causes in the process giving rise to that finite sequence, when this hypothesis is nevertheless true.

e_{22} We may accept the hypothesis that there existed, at the time the finite sequence was obtained, one or more assign-

[22] J. Neyman and E. S. Pearson, "On the use and interpretation of certain test criteria for purposes of statistical inference," *Biometrika*, vol. 28A, pp. 175–240, 1928; and in particular, p. 177. The italicizing in the quotation is mine. I have also introduced the symbols e_{11} and e_{21} instead of their numerals 1 and 2.

able causes in the process giving rise to that finite sequence, when this hypothesis is nevertheless false.

It should be noted that the hypothesis in this instance pertains to the existence of assignable causes during the time the finite sequence was being obtained.

The pair of errors e_1 and e_2, so far as they are encountered in interpreting the operation of control as a whole, may be stated similarly—

e_{13} We may reject the hypothesis that the production process or repetitive operation is in a state of statistical control when this hypothesis is nevertheless true.

e_{23} We may accept the hypothesis that the production process or repetitive operation is in a state of statistical control when this hypothesis is nevertheless false.

In this instance we should note that the hypothesis pertains to the conditions existing within a repetitive operation *throughout the time required to produce an infinite sequence.*

These three pairs of errors are alike in general form, but they differ in the hypotheses involved. They also differ in that Neyman and Pearson's errors e_{11} and e_{21} of the first and second kinds are essentially formal, whereas the other two pairs are expressed in empirical terms. For example, Neyman and Pearson can theoretically build an exact mathematical model that enables them to compute with any desired degree of exactness the probabilities of occurrence of their two kinds of errors. It will be noted that their hypothesis involves the assumption that the observed data constitute a *random* sample, and we have already considered some of the difficulties involved in trying to give this term an empirical and operationally verifiable meaning. In fact, we may think of the whole operation of statistical control as an attempt to give such meaning to the term random. But just as soon as we pass from the concept of the errors e_{11} and e_{21} of Neyman and Pearson, which may be defined in terms of a mathematical model, to errors of the general form e_1 and e_2 expressed in terms of scientific hypotheses about observable phenomena, we can no longer compute with mathematical exactness the probabilities associated with any pair of errors for a given hypothesis. As a background for the development of the operation of statistical control, the formal mathematical theory of testing a statistical hypothesis is of outstanding importance, but it would seem that *we must continually keep in mind the fundamental difference between the formal theory of testing a statistical hypothesis and the empirical testing of hypotheses employed in the operation of statistical control.* In the latter, one must also test the hypothesis that the sample of data was obtained under conditions that may be considered random.

The Judgment of Statistical Control

To form a background against which to view the problem of judging the condition of statistical control, let us summarize some of the points previously made. The engineer wants a product of uniform or homogeneous quality. As a basis for a quantitative characterization of such a product he conceives of one arising under a state of statistical control that assures (*a*) predictability in the probability sense and (*b*) minimum variability in quality. To attain this state, the engineer finds that he must go through certain operations of statistical control in which he uses a technique involving the use of statistical criteria for finding and weeding out assignable causes. The concept of a state of statistical control is a basis for describing the engineering goal of uniform quality, and the operation of statistical control is a means of approaching this goal. In any specific instance there remains the problem of judging how close one has approached the goal, and this is the problem now to be considered.

As a beginning, let us again consider the statement: "The quality of this product is in a state of statistical control" (see statement B on page 8). For our present purpose, we shall assume that this is equivalent to the statement that the quality of the product being turned out by the production process is uniform. Confining our attention to a single quality characteristic X, we may represent the quality of such a product by the infinite sequence

$$X_1, X_2, \cdots, X_i, \cdots, X_n, X_{n+1}, \cdots, X_{n+j}, \cdots \quad (3)$$

where the order in the sequence corresponds to the serial order in which the pieces of product are produced.[23] Let us consider the meaning of the statement that the quality X of this product is statistically controlled, remembering that at the time such a statement is made we have at our disposal only a finite number n of terms of the sequence.

We can draw three important conclusions. First, any such statement to be definite must be definite in respect to the meaning of the state of statistical control implied. Second, any such statement is a *probable inference* implying a prediction P about an *unobserved portion of the sequence.* Third, what we know about the n observed values of X and about the results obtained in applying the technique of statistical control to the production process constitutes the evidence E for the prediction P.

Let us refer to the time at which such a statement is made as the present, and let us assume that n terms of the potentially infinite sequence have

[23] For practical purposes of simplification, it is here tacitly assumed that the process or machine makes but one object at a time. In practice, of course, there is likely to be a whole battery of machines which may turn out more than one piece of product at a time. The treatment here given can be extended to cover this case, but would be unnecessarily involved for illustrating the fundamental points here considered.

been observed. Any such statement will be assumed to involve some kind of prediction about some portion or the whole of the unobserved part of the infinite sequence beginning with the term X_{n+1} as indicated schematically below:

$$X_1, X_2, \cdots, X_i, \cdots, X_n, \mid X_{n+1}, \cdots, X_{n+j}, \cdots$$
$$\text{Past} \qquad\qquad \text{Present} \qquad \text{Future}$$

For a prediction to have an operationally definite meaning, it is necessary that there be given or implied a perfectly definite way of determining whether it is true or false. Hence it is necessary that there be implied an operationally definite meaning of the statistical state of control in terms of characteristics of the sequence (3). There are two senses in which we may have such a meaning. One is the *theoretical* sense in which we include all possible criteria that the mathematical statistician may impose upon the infinite sequence (3) as a characterization of what he means by a mathematical state of control. The other is the *practical* sense in which one chooses a limited group of criteria to be applied in some specified way to a finite portion of the sequence consisting of $n + j$ terms, j of which have not been observed at the time the prediction is made.

Postulate II. In what follows we need to keep clearly in mind that the statement that the quality of product is in a state of statistical control involves a prediction P which may or may not be true, and it involves the evidence E for believing in the prediction. *The statement itself is a probable inference.* I shall assume the basic

> ***Postulate II.*** **The objective degree of rational belief p_b' in an inference involving a prediction P based upon evidence E is not an intrinsic property like *truth* but inheres in the inference through some relation of the prediction P to the evidence E.**

We can not here go into a discussion of all the reasons why it seems desirable to adopt this postulate, any more than earlier in this chapter we could go into a discussion of all the reasons for adopting Postulate I. It must suffice here to recall that in our discussion we have tried to show that in order to make successful use of probability we must consider a chain of three kinds of operations, viz., mental, physical, and mathematical. The fact that we must depend upon a human individual to choose successfully from his experience those conditions that he believes will lead to valid conclusions through the use of probability theory indicates what appears to be a necessary human act of rational believing and this act is always an attempt to relate past evidence E with a prediction P.

Three kinds of operations necessary in the successful use of probability—mental, physical, and mathematical

Starting with this postulate, we should note that the process of verifying an inference involving a prediction P based on evidence E must be different from that of verifying the prediction P. To verify the prediction, we do not need to have in mind any evidence, whereas to verify the inference we must determine whether or not the prediction P is reasonable upon evidence E. Thus it is obvious that *an inference based upon specified evidence E may be reasonable or valid upon the basis of that evidence even after one has learned that the prediction is false.*[24] These things will be discussed more in detail in chapter III

Now we are in a position to appreciate more fully the three concepts of statistical control (p. 1), namely, as a *state*, as an *operation*, and as a *judgment.* The state of statistical control is an ideal goal; statistical control as an operation is a means of attaining the goal; and concerning control there must be a judgment in the nature of a probable inference as to whether the state has been attained. The judge of quality must be familiar not only with the statistical means of specifying the state of statistical control in terms of which he makes his predictions but he also must be familiar with the rules of probable inference and rules of evidence. His job is in this sense closely analogous to that of the judge in the theory of Anglo-American jurisprudence; the legal judge has his rules of evidence and principles of judicial proof, and the judge of quality must have corresponding rules and principles, including those underlying statistical inference.

The judge of quality must be familiar with the rules of probable inference and rules of evidence

The Significance of Statistical Control

Let us first consider the significance of the operation for attaining and maintaining statistical control of quality upon statistical methodology. As we have tried to show in the discussion about the state of statistical control, there is a purely formal and mathematical theory of distribution which may be taken as characterizing our concept of a purely formal state of statistical control which, so far as the formal theory is concerned, may or may not be descriptive of any state attained or attainable in practice. Then there is the concept of the *physical* state of statistical control (drawings from the bowl universe), which represents the limit to which we can go in attaining valid predictability and minimum variability. Quality control studies have shown that there is good reason to believe that such a physical state can be attained in mass production, and that, when attained, the observables of this state satisfy the criteria that are used in describing the formal (mathematical) state.

[24] For a further and lucid discussion of this conclusion and related matters, see C. I. Lewis, *Mind and the World-Order* (Scribners, New York, 1929), pp. 309–344.

In the customary application of statistical theory, one assumes that he is dealing with a physical state that gives samples showing the characteristics of randomness. Control studies have shown that such physical states of statistical control are indeed rare natural occurrences, at least in physics and engineering (*cf.* later chapters), and furthermore that they can not usually be brought about without the operation of statistical control, wherein comparatively large numbers of preliminary data are taken in the process of detecting and removing assignable causes of variability. In this chapter, we have considered the problem of control only from the viewpoint of attaining valid predictability and minimum variability in a measured quality X. In other words, we have neglected the matter of accuracy, which will be considered later, especially in chapter IV. We shall then find still more evidence to indicate the need for going through a definite operation to attain a state of statistical control before applying statistical theory that is based on the assumption that such a state exists.

Formal distribution theory will give valid predictions only in a state of statistical control

Next let us consider the significance of the study of statistical control from the viewpoint of the control of quality. Let us recall the three steps of control: specification, production, and judgment of quality (page 1). On the older concept of an exact science these three steps (call them I, II, and III) would be independent. One could specify what he wanted, some one else could take this specification as a guide and make the thing, and an inspector or quality judge could measure the thing to see if it met specifications. A beautifully simple picture!

The whole picture, however, is radically different just as soon as we admit that we have only a probable science. Even when we limit ourselves to trying to stay within tolerance limits, it is necessary for economic reasons and for attaining maximum quality assurance in all kinds of work, including that where tests are destructive, to introduce the concept of action limits A and B and the aimed-at value C, fig. 6. But in order to specify C we must first apply the operation of statistical control. In fact the C must really come from Step III and after suitable action limits A and B have been established in Step II. But these action limits can not be set without some knowledge of the tolerance limits that are specified in Step I. I think it is particularly important to note that the third step can not be taken by simply inspecting the quality of the objects as objects, but instead must be taken by inspecting the objects in a sequence ordered in relation to the production process. In fact these three steps must go in a circle instead of in a straight line, as shown schematically in fig. 10. It may be helpful to think of the three steps in the mass production process as steps in the scientific method. In this sense, specification, production,

Specification, production, inspection (p. 1) not independent

and inspection correspond respectively to making a hypothesis, carrying out an experiment, and testing the hypothesis. The three steps constitute a dynamic scientific process of acquiring knowledge. From this viewpoint, it might be better to show them as forming a sort of spiral gradually approaching a circular path which would represent the idealized case where no evidence is found in Step III to indicate a need for changing the specification (or scientific hypothesis) no matter how many times we repeat the three steps. Mass production viewed in this way constitutes a continuing and self-corrective method for making the most efficient use of raw and fabricated materials.

The three steps in fig. 10 correspond to the three steps in a dynamic scientific process of acquiring knowledge

From the viewpoint of specification it is of interest to note that for the meaning of control to be operationally definite, not only certain criteria of control, but also the operation of selecting the objects whose qualities are to be tested, must be specified. The choice of criteria to be used as a method

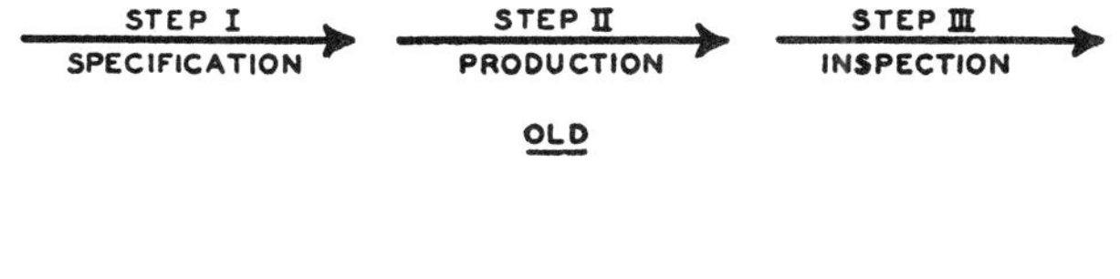

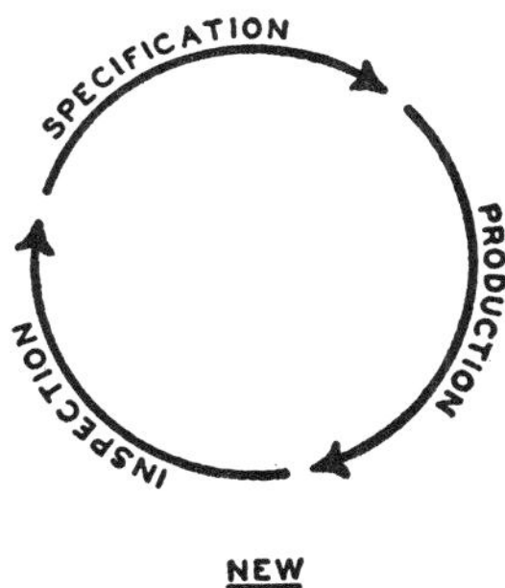

FIG. 10

of verifying the state of control can be made without reference to a given kind of product, but the method of specifying the sequence to be used in the chosen criteria can not in general be set down without reference to empirical information obtained in production. What is still more important, the intent of any such specification implies a certain degree of assurance that the quality of the product will be found to satisfy this set of criteria, particularly when not every piece of the product can be tested. Here again, without a knowledge of the results of prior attempts to control quality, one can not specify in a perfectly definite way just how many data are required and in what sequence these data shall be used in applying control criteria to give the quality assurance intended by the design specification. For these reasons it seems that operationally verifiable control requirements, and requirements as to how many data shall be obtained to provide adequate quality assurance, can only be set down in Step III, and then only by one having his eye both on the intent of design requirements and upon the accumulated inspection results to date, indicating the degree to which a state of statistical control has been approached. Hence the design specification must be supplemented in Step III by inspection practices providing adequate data and satisfactory criteria of control for each type of product.

Furthermore, since the running record of past results must play such an important part in judging the degree to which control has been attained, it is necessary that Step III provide such a continuing record or quality report. The graphical control chart (Criterion I) is admirably adapted to this end. The mathematical theory of distribution characterizing the formal and mathematical concept of a state of statistical control constitutes an unlimited storehouse of helpful suggestions from which practical criteria of control must be chosen, and the general theory of testing statistical hypotheses must serve as a background to guide the choice of methods of making a running quality report that will give the maximum service as time goes on.

Statistical theory and techniques enter each of the three steps of fig. 10

To attain economic control and maximum quality assurance, statistical theory and techniques must enter every one of the three steps in the control of quality. In this way they make possible a very important potential contribution of mass production to scientific industrial progress. Incidentally, we have seen that this potential state of economic control can be approached only as a statistical limit even after the assignable causes of variability have been detected and removed. *Control of this kind can not be reached in a day. It can not be reached in the production of a product in which only a few pieces are manufactured. It can, however, be approached scientifically in a continuing mass production.*

The Future of Statistics in Mass Production [25]

Much has been written about the application of statistical theory and techniques in studying, discovering, and measuring the effects of an existing system of unknown or chance causes. Much remains to be written about the application of statistical theory and techniques in finding out how to

[25] Extracted from a paper by this title delivered at the Detroit meeting of the American Institute of Mathematical Statistics, December 1938.

tinker with and modify an existing chance cause system until it behaves as we want it to. The statistician knows that his predictions will be valid if certain assumptions about the cause system are justified, perhaps the most important assumption being that the particular effects of his chance cause system are random. In mass production the statistician has learned by experience that random effects do not just happen, even by careful planning. If the industrial statistician ignores this fact and makes predictions as if he were dealing with randomness, he may expect many of his predictions to go wild; what is more he knows that this fact will be discovered and his work discredited. For this reason the industrial statistician in mass production must commence by developing techniques for determining when we are justified in assuming that the effects of the underlying cause system are random, and when the usual distribution theory is applicable.

Experience in the control of quality has provided a practical technique for detecting and eliminating assignable causes of variability in the production process until a state of statistical control is reached wherein predictions based upon the assumption of randomness will prove valid. By the elimination of assignable causes of variability we make the most efficient use of raw materials, maximize the assurance of the quality of the manufactured product, minimize the cost of inspection, and minimize loss from rejections. Statistics in mass production can be made to pay good dividends, and has a bright future. What does this future depend on?

The answer to this question is contained in the three fundamental steps in quality control (p. 1; also fig. 10, p. 45):

I. The specification of the quality of the thing wanted.
II. The production of things designed to meet the specification.
III. The inspection of the things produced to see whether they meet the specification.

We have seen that the outstanding characteristic of the first step is the necessity of setting up and putting into effect a tolerance range for each specified quality characteristic. If a producer contracts to deliver goods within some specified range and upon applying Steps II and III finds that some of his product falls outside the tolerance limits, he loses money. He must not contract to meet tolerance limits that are too narrow, yet if he is to make the most efficient use of materials, he must, in most instances, close up the tolerance limits as much as he dares.

Obviously one can not specify a practically attainable tolerance range out of thin air; one must recognize what is possible under commercial conditions of production in Step II, which in turn is revealed by inspection in

Step III. He must also take into account the fact previously noted, and on which much more will be said later, that the manufacturing process to begin with is almost certain not to be in a state of statistical control. This state can be approached only through the application of certain statistical techniques involving the use of the control chart. The point to be stressed is that the three steps, specification, production, and inspection, can not be taken independently in mass production: instead they must be coordinated, each step being of assistance toward the attainment of the other two, as is suggested in fig. 10 (p. 45). In fact, the three steps may be thought of as a scientific experiment in which the objective is the attainment of the most efficient use of the available materials.

Broadly speaking, the statistician of the future has before him the opportunity of helping to develop this fundamental type of experiment. As has been stated, he must start by designing statistical control techniques for the elimination of assignable causes of variability, whereupon he can use modern statistical theories as described in the literature with reasonable assurance that his predictions will be found valid. He must, however, go further than is customarily recognized in the current literature in that he must provide operationally verifiable meanings for his statistical terms such as random variable, accuracy, precision, true value, probability, degree of rational belief, and the like. The chapters that follow will be an initial step in this direction.

In one sense the statistician's problem in mass production is more complicated than the design of experiments that is usually considered in the literature of statistics. Whereas the customary statistical theory is concerned with comparatively small-scale experiments carried out under laboratory conditions by a few people, the corresponding development of the mass production process must be carried out under commercial conditions on a large scale, involving large numbers of people. To illustrate, the three steps in the mass production process are usually carried out either by different companies or by different departments of the same company. The steps may involve the coordinated effort of literally hundreds and even thousands of employees, including physicists, chemists, engineers, sales agents, purchasing agents, lawyers, and economists. Very few of these people have ever had any training in statistics or probability, and yet they must be brought to appreciate them if the statistician is to develop the opportunity of making his full contribution. This situation constitutes a problem not only for those now in industry but also for those responsible for the training of the industrial leaders of tomorrow so that they will have sufficient knowledge of statistics to be able to recognize the potential contributions that statistical theory and technique have to offer.

In the future the statistician in mass production must do more than simply study, discover, and measure the effects of existing chance cause systems: he must devise means for modifying these cause systems to bring about the results that are desirable in the most efficient use of materials. He must not be satisfied simply to measure the demand for goods; he must help to change that demand by showing, among other things, how to close up the tolerance range and to improve the quality of goods. He must not be content simply to measure production costs; he must help to decrease them.

An additional duty for the statistician in mass production

The future contribution of statistics in mass production lies not so much in solving the problems that are usual to the statistician today as in taking a hand in helping to coordinate the steps of specification, production, and inspection. The long-range contribution of statistics depends not so much upon getting a lot of highly trained statisticians into industry as it does on creating a statistically minded generation of physicists, chemists, engineers, and others who will in any way have a hand in developing and directing the production processes of tomorrow.

CHAPTER II

HOW ESTABLISH LIMITS OF VARIABILITY?

> **Thus in many directions the engineer of the future, in my judgment, must of necessity deal with a much more certain and more intimate knowledge of the materials with which he works than we have been wont to deal with in the past. As a result of this more intimate knowledge his structures will be more refined and his factors of safety in many directions are bound to be less because the old elements of uncertainty will have in large measure disappeared.[1]**
>
> **FRANK B. JEWETT, *President***
> ***Bell Telephone Laboratories, Inc.***

WHAT IS INVOLVED IN THE PROBLEM?

In the previous chapter we saw how the engineer first tried to make things exactly alike in the process of mass production; how, for economic reasons, he was forced to adopt the use of the go tolerance limit and then the go, no-go tolerance limits; and finally how he was forced to adopt the use of the go, no-go tolerance limits *plus* two action or control limits and a statistical limit in order to effect additional economies and to attain maximum quality assurance. Attainment of the state of statistical control considered in the previous chapter involves the establishment of the control and statistical limits. The problem considered in this chapter is that of establishing the tolerance limits.[2] That is, we shall consider the question, how is the engineer of the future going to provide himself with a knowledge of the properties of materials that is adequate for setting tolerances in a way to make the most efficient use of these materials?

Note on the meaning of tolerance limits. Probabilities involved. We may think of the use of the go, no-go tolerance limits as constituting a means of *screening* a given product in respect to some quality characteristic. In this sense, tolerance limits on a quality characteristic X fix the range within which the quality X of a piece of product must lie in order to conform to specification and to fit into some mechanism that the engineer wants to make. From this viewpoint, the choice of limits depends upon a particular design. However, it is not only what the engineer wants but *what he can get*, or at least what he can get economically, that must be taken into account in

[1] "Problems of the engineer," *Science*, vol. 75, pp. 251–256, 1932.

[2] The relations between the five limits, two tolerance limits L_1 and L_2, two action or control limits A and B, and the statistical limit C, is illustrated in fig. 6, p. 24.

the setting of tolerance limits. So soon as an engineer undertakes to set tolerances that make efficient use of materials, he must think not only of the tolerance range itself but also of *the percentage of the product made under commercial conditions that may be expected to have a quality falling within this range.* Hence the establishment of *economic* tolerance limits necessitates the acquisition of knowledge concerning the probability that the product made under commercial conditions will have a quality falling within these limits.

Knowledge of the product under commercial conditions necessary when thinking of tolerance limits

There is another reason why the engineer under certain conditions must be concerned not only with the tolerance range but also with the probability associated with that range. For example, if the inspection test to determine whether the quality of a piece of product lies within the specified tolerance range is destructive, then it is only through a knowledge of the expected variability of quality that an engineer can determine what assurance he has that the quality lies within its tolerance limits.

With tolerance limits there must be an associated probability

Whereas tolerance is sometimes defined either as the difference between two limiting sizes as a means of specifying the degree of accuracy or as a specified allowance for variations from a standard, the concept of tolerance as used in this monograph implies not only the concept of tolerance limits *but also that of the percentage of the commercial product that may be expected to have a quality falling within this tolerance range.* So long as we think of a tolerance range simply as go, no-go limits, our attention is centered primarily on the limits themselves. However, just as soon as we begin to consider the establishment of tolerance limits from the viewpoint either of making efficient use of available materials or of maintaining an adequate degree of quality assurance, especially when the inspection test is destructive, we must think not only of the tolerance limits but also of the probability associated with these limits.

Three typical tolerance ranges. Let us confine our attention to a single quality characteristic X. Three typical kinds of tolerances that arise in practice are illustrated schematically in fig. 11. If p represents the probability [3] of a value of X falling outside the tolerance range L_1, L_2, the problem may be thought of as that of setting tolerance limits in such a way that

$$p \lesssim p' \tag{7}$$

where p' represents the largest fraction nonconforming that is allowable from an economic viewpoint. Associated with any such requirement there

[3] Some difficulty is involved in interpreting the meaning of this probability when the quality of the product is not in a state of statistical control because there is no *constant* probability p under these conditions: the probability p itself then varies with time.

is some tolerance range L_1, L_2, and it is desirable in most instances that this range be reduced to an economic minimum through the process of eliminating assignable causes of variability.

Very often in practice only the tolerance range is specified. As an example of a tolerance range in which both limits are specified we have the requirement that the diameter of a shaft must lie within the range L_1, L_2

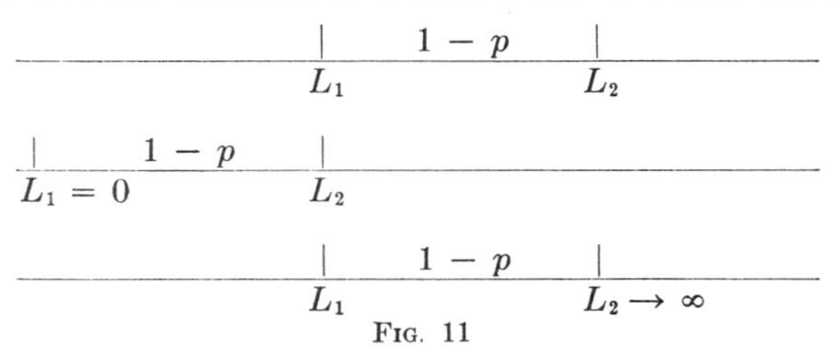

FIG. 11

(top of fig. 11). As an example of a tolerance range in which only the upper limit is specified, we have the requirement that the blowing time of a fuse shall not be greater than L_2 seconds (middle of fig. 11). By implication the lower limit is $L_1 = 0$. As an example of a tolerance range in which only the lower limit is specified, we have the requirement that the tensile strength of a steel strand shall not be less than L_1 pounds per square inch (bottom of fig. 11). By implication the upper limit is $L_2 = \infty$. Even though no requirement such as (7) is explicitly stated in any one of these three illustrations typifying practice, some such requirement is implied, because it is essential that the fraction p of nonconforming pieces shall not exceed some value that is usually less than 1 percent and often less than 0.1 percent.

Object of the chapter

Our object in this chapter may now be more definitely stated as that of trying to determine some of the potential contributions and inherent limitations of the application of statistical theory in the establishment of the *economic* tolerance limits L_1 and L_2 in each of the three cases.

THE PROBLEM FROM THE VIEWPOINT OF STATISTICAL THEORY

In order to see clearly the role that statistical theory may be expected to play in making possible the most efficient use of engineering materials, it is necessary to consider two fundamentally different conditions under which we are called upon to establish economic tolerances; namely, setting tolerances on the quality characteristics of (a) raw and fabricated materials and pieceparts, and (b) the completed unit, physical system, or engineering structure. For the sake of simplicity, let us think of some quality characteristic X of some fabricated material such as the tensile strength of steel, or that of malleable iron, the thickness of condenser paper, or the like.

First of all there is the problem of discovering the way the given property X varies under commercial conditions of production and then comes the problem of making the best use of material with this kind of variation when used in the design of complicated structures.

It is helpful to think of these two problems as being characteristically *inductive* and *deductive* respectively. The first has the earmarks of the so-called statistical problems of estimation and the second has the earmarks of the statistical problems of distribution.[4]

From an engineering viewpoint, these two problems, broadly speaking, may be considered as belonging in the field of research on the quality of materials on the one hand, and in that of design on the other. It is significant for both the engineer and the statistician that statistical theory can be made to play an important role in these two fields. For the engineer it makes certain economies possible and provides a rational basis for establishing interrelated tolerances in a complicated structure so as to make the most efficient use of materials. For the statistician it opens up a new field for the application of statistical theory and techniques not only in the inductive process of adding to our present knowledge of physical properties of materials and physical laws but also in the *deductive process of designing* structures that make the most efficient use of our present knowledge of available raw and fabricated materials.

If it were possible to attain a state of statistical control for each and every quality characteristic of fabricated materials and if the frequency distributions of these characteristics were known, it is obvious that the efficient use of such knowledge in designing new structures would involve, among other things, the direct application of mathematical distribution theory. Even before the engineer has attained the state of statistical control for all important quality characteristics of the fabricated materials and pieceparts entering into a given design, it is possible under certain conditions to effect a reduction in overall tolerance limits and to decrease the number of rejections by *randomizing* the assembly process so as to distribute the effects of assignable causes. The applications of statistical theory in the processes of design and assembly are particularly attractive to the mathematical statistician for they are likely to pay a good return on the exercise of all his mathematical talents. However, such applications are only touched upon here and there in this monograph. In the rest of this chapter, for example, we shall consider primarily the inductive process of establishing tolerances on a single quality characteristic X of any fabricated material or piecepart.

[4] Of course, when making this distinction, we must keep in mind that inductive scientific inference involves the use of both inductive and deductive steps employed in the process of making and testing a hypothesis.

A practical example. One might expect that all the engineer needs to do in order to improve his technique in setting tolerances on some quality characteristic X is to become acquainted with the available theory of "statistical estimation." We shall later find that such an expectation is not justified, but that is getting ahead of our story.

Let us assume that we wish to use malleable iron in some design and to set tolerance limits on its tensile strength. We naturally turn to the engineering literature for data obtained under practical conditions to be used as a basis for setting such tolerances. In the report of a recent symposium [5] on malleable iron, we find the results of 5000 tensile-strength measurements on as many test bars of this material. These measurements were made by Enrique Touceda for the Malleable Iron Research Institute, the bars having been taken from several different heats over the period from May to November 1930, from each of the companies comprising the membership of the Institute. These data are presented in table 1. Here we

TABLE 1

TENSILE STRENGTH OF 5000 MALLEABLE IRON BARS

Range of values lbs. per sq. in.	*Observed distribution*	*Normal law distribution*	*Difference*
Under 45,000	0	0	0
45,000–45,999	1	0	1
46,000–46,999	2	1	1
47,000–47,999	3	5	−2
48,000–48,999	8	22	−14
49,000–49,999	23	77	−54
50,000–50,999	289	210	79
51,000–51,999	472	447	25
52,000–52,999	739	744	−5
53,000–53,999	927	963	−36
54,000–54,999	967	970	−3
55,000–55,999	758	762	−4
56,000–56,999	481	466	15
57,000–57,999	230	222	8
58,000–58,999	72	82	−10
59,000–59,999	19	24	−5
60,000 and over	9	5	4

have a very respectable-looking unimodal frequency distribution (fig. 12). However, when graduated to a normal curve, the closeness of fit between the theoretical and the observed distributions as measured by $\chi^2 = 90.23$ is not very good, and the theoretical statistician might therefore argue that the hypothetical universe is not normal. For our present purpose, however, the value of χ^2 is of no interest; *we are not concerned with the functional form of the universe but merely with the assumption that a universe exists.* If it exists at all, then it would appear that the setting of economic tolerances

[5] Symposium on malleable iron castings, published in the *Proc. Amer. Soc. Testing Materials*, vol. 31, pp. 317–434, 1931.

reduces to a statistical problem of estimation; and by increasing the sample size at will, we could presumably approach closer and closer to the tolerance

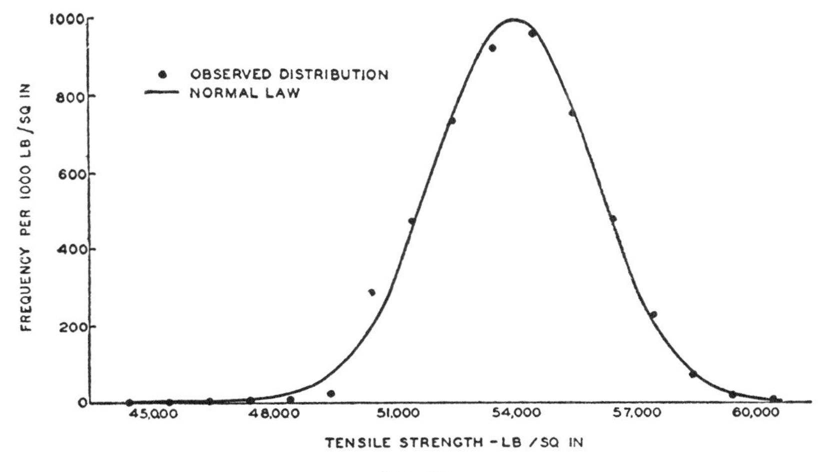

FIG. 12

range associated with any specified value of probability.[6] If, however, the universe does not exist, there is no corresponding rule for getting closer and closer to the tolerance range. Our problem here of setting tolerances is therefore twofold: (*a*) to examine the available evidence to see if one is justified in assuming that a statistical universe exists, and (*b*) to consider the technique of setting tolerances both when the assumption is justified and when it is not.

Setting tolerances is a twofold problem

There is, however, another aspect of setting tolerances that we must consider. For example, let us assume that we wish to make use of pure iron in some way that requires us to set tolerances on its density. Accordingly we turn to an authoritative table [7] of physical properties and find the density given as (7.871 ± 0.002) gms/cm^3. This example is typical of the case where the available information upon which to base tolerances is given in the form $X \pm \Delta X$. What is the meaning of such a range and what relation, if any, does it bear to the tolerance range? Obviously, a tolerance range can be put into this form so far as its numerical aspects are concerned.

[6] For a more critical discussion of the limiting process here involved, see the latter part of chapter IV.

[7] *Physical Constants of Pure Metals*, The National Physical Laboratory (His Majesty's Stationery Office, London, W. C. 1, 1936).

Furthermore, if we turn to the literature of modern statistics, we find much emphasis placed upon the assertion that with the help of modern small-sample theory such ranges can be established upon the basis of small samples just as validly as upon the basis of large samples. Hence the engineer rightly wants to know if the statisticians have found a satisfactory way for setting tolerance ranges on the basis of small samples.

Can valid tolerance ranges be fixed on the basis of small samples?

We find much confusion regarding the meaning of ranges $X \pm \Delta X$ even in the literature of statistics. The fact that the meaning of the range that is valid in the sense of so-called modern small-sample theory turns out to be different from the meaning of the tolerance range should be of considerable interest to statisticians as such as well as to engineers.

How Establish Tolerance Limits in the Simplest Case?

A tolerance range for the bowl universe. Instead of tackling at this point the practical problem of setting tolerance limits on a property such as the tensile strength of malleable iron castings, let us start with the simpler problem of establishing tolerance limits where we *know* that the sample $X_1, X_2, \cdots, X_i, \cdots, X_n$ of data was drawn one at a time with replacement from an experimental [8] normal universe. Let us consider first how to set a tolerance range $X = L_1$ to $X = L_2$ that will include let us say $(1 - p')N = .5N$ or one-half of N future drawings from the bowl. An engineer may wonder why we choose .5 whereas in practice p' is most likely to be less than .01: we choose this value of p' because several books in science and in error theory seem to tell one just how to establish L_1 and L_2 for $p' = .5$. For example, one outstanding treatise of 1937 on a particular branch of physics has an appendix discussing accuracy and precision. The authors give eleven measurements of a length. They calculate the arithmetic mean $\bar{X}$ of this sample and the estimated probable error e of a single measurement in accord with classical error theory. They then state in effect that if another set of n measurements be made under the same conditions, it is an even chance that the mean of this set will differ from the mean of the set of eleven measurements by more than $e/\sqrt{n}$. This certainly looks to the uninitiated like a means of setting a tolerance range for a probability of $\frac{1}{2}$. Of course, the implication is that ranges for any probability could be set up in an analogous manner with proper allowance for the magnitude of the desired value of p', such as 0.1, 0.05, etc.

Sometimes the result of an experiment is more convincing than an argument, and therefore let us see what might happen to one who set tolerance

[8] See, for example, page 165, table 22, of my *Economic Control of Quality of Manufactured Product*, for such a distribution.

ranges by such a rule. For this purpose, I drew from a normal universe in a bowl the sample of eleven measurements shown in table 2. The average $\bar{X}$, and the estimated probable error e of a single observation are .009 and .322 respectively. Now let us set up tolerance limits for a probability of $\frac{1}{2}$ and sample size [9] $n = 4$. In line with the previous paragraph, we find that such limits would be $.009 \pm .161$, since $.322/\sqrt{4} = .161$. According to the

TABLE 2

.5	.1	−.3	−.9	.1	−.1
−.1	.3	−.6	.7	.4	

$\bar{X} = 0.009$, S.D. $= 0.456 = \sigma$, $0.674\sigma\sqrt{11/(11-1)} = 0.322 = e$

authors of the text, at least as I interpret their discussion, we should expect to find fifty percent of the averages of samples of four lying within this range. Well, let us take 100 samples of four and see if such a prediction is valid. Fig. 13 shows the results of one such test. We were led to expect 50 percent within the limits $.009 \pm .161$ shown by the dotted lines: actually we find 27 percent! The prediction of 50 percent within limits was not valid!

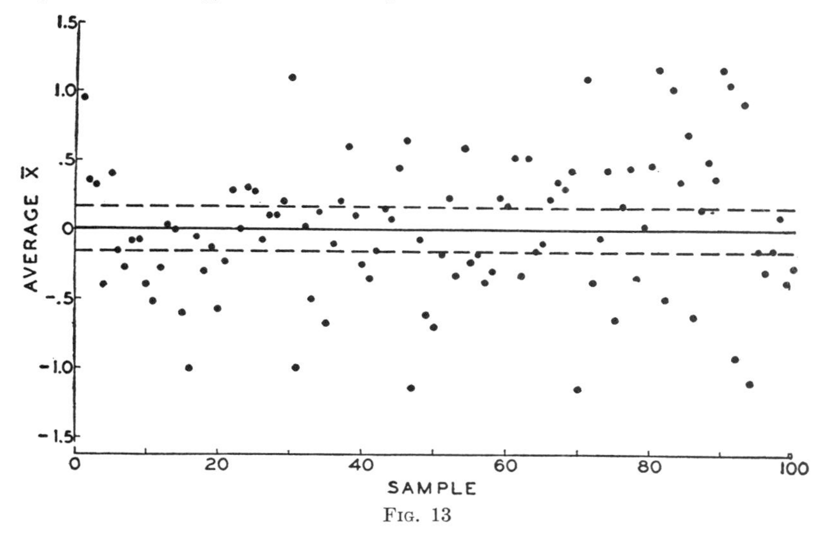

FIG. 13

What would happen to a practical man who followed such a rule? In answering, I am reminded of the old saying: when a doctor makes a mistake, he buries it; when a judge makes a mistake, it becomes the law. I would add in the same vein: when a scientist makes a mistake in the use of statistical theory, it becomes a part of "scientific law"; but when an industrial statistician makes such a mistake, woe unto him for he is sure to be found

[9] Of course, I might have chosen any other value of n.

out and get into trouble. Why the difference? The answer is that in establishing tolerances, one can rest assured that he will hear about it if appreciably more than the expected percentage of the product is found outside of the limits, because hundreds, thousands, and sometimes even millions of pieces of product are made per month.

It would, of course, be unfair for the engineer to judge the usefulness of statistical theory on the basis of the example just considered. Gauss, the originator of the estimate $\sigma\sqrt{n/(n-1)}$, was aware of the fact that it fluctuates from sample to sample, and the same can be said of all careful writers since his time. Some of the inherent limitations of the older theory have been overcome by work that began in 1908 with Student's publication [10] of tables for the probability p_z that the mean of a sample of n, drawn at random from a normal population, will not differ from the expected value $\bar{X}'$ of the population by more than z times the standard deviation σ of the sample. Let us see whether this fundamental contribution helps us to set tolerance limits for the ideal case of the normal bowl universe.

Student's theory

First, let us see just what this theory enables us to predict with validity. Interpreted in an operationally verifiable way, this theory means, among other things, that given a normal universe, even though its expected value $\bar{X}'$ and standard deviation σ' are unknown, we can nevertheless make the valid prediction that if we draw a series of N samples of size n, and calculate the N ranges [11]

$$\bar{X}_1 \pm z\sigma_1, \qquad \bar{X}_2 \pm z\sigma_2, \cdots, \qquad \bar{X}_N \pm z\sigma_N$$

then p_zN of these ranges may be expected to include the expected value $\bar{X}'$ of the population. If the population is an experimental one whereof the theoretical limiting value $\bar{X}'$ may be obtained,[12] then such a prediction can be tested. As an example, fig. 14 shows a series of 100 such ranges for $n = 4$, 40 ranges for $n = 100$, and 4 ranges for $n = 1000$. The ranges were all calculated with $p_z = \frac{1}{2}$. The expected value [13] $\bar{X}'$ is zero, and is shown

[10] Student, "The probable error of a mean," *Biometrika*, vol. 6, pp. 1–25, 1908. As Birge once remarked to the editor, Student nowhere in this paper mentioned the probable error of a mean except in the title.

[11] For samples of 4, and with $p_z = \frac{1}{2}$, $z = 0.442$, as is found from Student's integral or Fisher's table of t, the relation being $t = z\sqrt{(n-1)}$. For samples of $n = 100$ or $n = 1000$, one may take $z = 0.6745/\sqrt{(n-1)}$, since at such large values of n the normal and Student integrals are practically equal for abscissas not too far into the tails.

[12] For a critical discussion of the operational sense in which a theoretically true value such as $\bar{X}'$ can be obtained, see the discussion of logical verifiability and of the meaning of *concepts in use* in the latter part of chapter IV.

[13] Of course, zero is the average of the numbers on the chips in the bowl, yet the statistical limit $\bar{X}'$ of $\bar{X}$ may not be zero. We have no way of telling, but an operational method of circumventing this difficulty will be discussed later on in this chapter (see sequences (11) and eqs. (12) and attendant discussion), also more fully in the last chapter. *Editor.*

by the heavy central line. By actual count, the percentages of ranges that include zero are 51, 45, and 50, the expected value p_z being 50. To me, this constitutes an excellent check between theory and experiment. Thus we

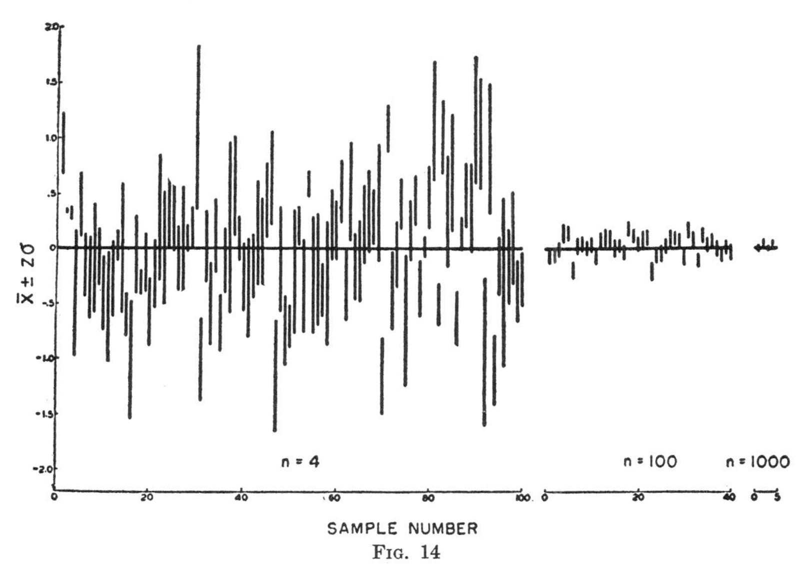

FIG. 14

see that it is possible to make predictions with Student's theory regarding the *varying* ranges in the sense of fig. 14 that are just as valid for small samples as for large ones.[14]

[14] *Editor's comment.* It is important to realize that the change from bad agreement to good agreement that has been brought about by substituting the Student ranges in fig. 14 for the single range in fig. 13 is due more to a development in interpretation than to any numerical refinement provided by Student's integral.

In fig. 13 there is only one range—the spacing of the horizontal lines; moreover this one range is centered throughout at 0.009, which happened to be the mean of the 11 measurements on page 57. This one range just happened to give bad results; it might have been somewhat wider and given better results; if it had accidentally been close to $2 \times 0.674\sigma'/\sqrt{4}$, spaced centrally about 0 (.009 is close enough), it would have worked very well. In fig. 14 there is not a single range, but many ranges—one for each sample, their centers and lengths following the fluctuations of the means and standard deviations of the successive samples. This would be so whether each range had been computed with the classical estimate of the probable error made from that particular sample ($.674\sigma/\sqrt{(n-1)} = .389\sigma$ for samples of 4), or with Student's multiplier ($.442\sigma$), as was actually done. Of course the latter will give somewhat better results with the normal bowl, on the average, but the numerical refinement of replacing .389 by .442 is not so momentous as has been proclaimed by many writers. Of much more importance to the statistician is the fact that, whether he uses the classical estimate $\sigma\sqrt{n/(n-1)}$ of σ', or Student's integral, he is at the mercy of the sampling fluctuations of σ, even in controlled experiments.

The chief lesson in figs. 13 and 14 is the resignation to the fact that no single small sample can provide the information needed for setting the width of a single pair

Student's theory inadequate for tolerance limits. Now let us consider the problem of establishing a tolerance range for samples from the normal bowl that was used in getting the data shown in figs. 13 and 14. As before, let us assume that we do not know the parameters of the normal distribution in the bowl. Our problem is to set a range $X = L_1$ to $X = L_2$ such that the probability of drawing a value X lying within this range is some previously specified value, $1 - p'$. As a special case let us take $p' = .5$. We shall assume that the only way we can find out anything about the normal universe in the bowl is by making drawings with replacement.

Obviously the starting point is to draw a sample of n values of X from the bowl. To make the problem specific, let us assume that we have drawn the following sample of four:

$$1.7, \quad 0.2, \quad 1.4, \quad 0.5$$

How shall we set up the tolerance limits L_1 and L_2 for a probability $p' = .5$?

I think it will be generally (perhaps unanimously) agreed among statisticians that our best estimate of such a range can be put in the general form $\bar{X} \pm t\sigma$. It is obvious, however, that no matter what rule is adopted for computing such a range, that range will only as a rare event correspond to a probability $p' = .5$. It is also obvious that the problem of establishing a valid tolerance range is *fundamentally different from the problem solved by Student.* His theory tells how to make valid predictions of the number of times a series of *varying ranges* with *varying centers* may be expected to include a theoretically true value, whereas, in order to establish a valid tolerance range, we must be able to make a valid prediction about how many times future observed values may be expected to fall within a given pair of *fixed limits.*

A study of three types of ranges. The difference between the Student type of range and the estimated tolerance range is of fundamental importance. The two ranges should certainly not be confused as they sometimes

of lines that will perform the feat that was expected of those in fig. 13, and the substitution of a theory that deals with the *varying* ranges of fig. 14. It should be noted that any prediction involving a Student range $\bar{X} \pm z\sigma$ is a probability prediction concerning not that particular range, but rather a *whole sequence of varying ranges.*

In control work, and also, I fear, in much of the application of statistics to agriculture, we need to get back to the idea of a *single range*—a pair of horizontal lines like those in fig. 13—but *they must be in the right place.* The fluctuating ranges of fig. 14 will not suffice, even though close to 50 percent of them in a long series do overlap the true value, just as predicted by Student's theory.

The editor desires to point out that though there was a prior publication of a chart similar to fig. 14 in Deming and Birge's *Statistical Theory of Errors* (their fig. 11), the notion came originally from conversations with Shewhart, as explained in Deming and Birge's footnote 27. An illuminating chart for illustrating the distinction between the samples included by Student's integral and those included by the normal integral is fig. 4 on page 341 of an article by Alan Treloar and Marian Wilder, *Annals of Mathematical Statistics,* vol. 5, pp. 324–341, 1934.

are in the literature, particularly when the probability is taken as .5. Neither should the meaning of either of these ranges for a probability of .5 be confused with the meaning of the probable error range $\bar{X}' \pm .6745\sigma'/\sqrt{n}$ of classic error theory for averages of samples of size n, where $\bar{X}'$ and σ' are the true average and standard deviation respectively of the universe. To emphasize this point let us consider the following examples of these three ranges:[15]

50 percent Student range:[16]	$\bar{X} \pm z\sigma$ computed from a sample of n in such a way that 50 percent of the ranges computed in this same way from a sequence of samples of n from this same universe may be expected to include $\bar{X}'$.
Estimated 50 percent tolerance range:[17]	$\bar{X} \pm k\sigma$ computed from a sample of n and estimated to be the range that will include 50 percent of the averages of future samples of n from the same universe.
The probable error, or 50 percent tolerance range:[18]	$\bar{X}' \pm .674\sigma'/\sqrt{n}$ computed from the universe parameters and assumed to include 50 percent of the averages of future samples of n from the same universe.

The following five points should be noted: (1) the first two ranges are computed from a sample, whereas the third is computed from the parameters of the universe; (2) the predictions involved in the meanings for the second and third ranges are the same, and this prediction is different [19] from the corresponding prediction for the Student type of range; (3) the validity of the prediction for either the Student or the probable error range is to be independent of the sample size, whereas the validity of the *estimated* tolerance range (the second one of the three) depends upon the sample size n in a way that we shall shortly consider in some detail; (4) as the sample size n in the estimate of the tolerance range approaches infinity, we may expect

[15] Probabilities of other than 50 percent could be considered with obvious modifications in the ranges. In fact, in the example treated in fig. 15, the fraction used is 99.73 percent.

[16] The 50 percent Student range is what Neyman and Pearson would call the "50 percent confidence interval" computed from the standard deviation of the sample. Here the 90, 95, 99 percent confidence intervals would be computed similarly, but with different values of z, as found by Student's integral. Values of z are conveniently found from Fisher's table of t, the connection being $t = z\sqrt{(n-1)}$. Deming and Birge show the 50 percent values of z directly in their *Statistical Theory of Errors*, p. 140. *Editor.*

[17] It should be noted that the z and k are different numbers even for the same probability because the ranges $\bar{X} \pm z\sigma$ and $\bar{X} \pm k\sigma$ are not subject to the same interpretation.

[18] In chapter IV it is pointed out that there is no possible *physically operational meaning* to this third type of range.

[19] See editor's comment on pp. 59–60.

predictions in terms of these estimates to approach the same degree of validity in the statistical sense as predictions in terms of probable error ranges; and (5) as the sample size used in computing either the Student or the estimated probable error range approaches infinity, both these ranges approach in the statistical sense the probable error range computed from the parameters of the universe, as given above.[20]

Now let us see what we must do in order to set up a tolerance range for a prediction which is valid within limits that are practical. For this purpose let us choose $1 - p' = .9973$ because this is about the magnitude customarily used in engineering practice. Of course, if we knew $\bar{X}'$ and σ', the desired range would be $\bar{X}' \pm 3\sigma'$. Let us see what happens if we take $\bar{X} \pm 3\sigma\sqrt{n/(n-1)}$ as the range for each sample. Fig. 15 shows 100 such ranges for as many samples of 4 drawn from an experimental universe; 40 ranges for 40 samples of 100; and 4 ranges for 4 samples of 1000. The dotted limits are $\bar{X}' \pm 3\sigma'$.

A tale of great practical importance hangs on this figure. The standard deviation σ fluctuates from sample to sample so wildly for samples of four that large errors in prediction often result. But for n so large as 100 the standard deviation is much steadier, and for $n = 1000$, steadier yet. If one were to go through life setting 99.73 percent tolerance ranges for samples of four,

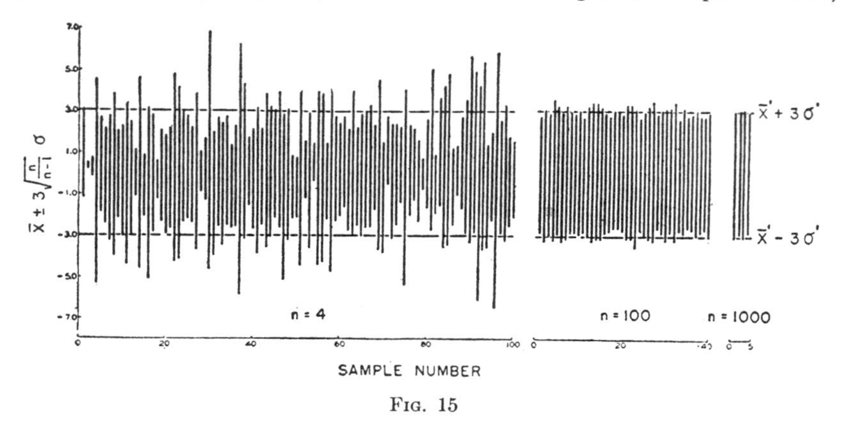

FIG. 15

using the "estimated" value of σ' as indicated in the previous paragraph, he would sometimes get a range that includes a very small percentage, even when the samples are drawn from a normal universe. For example, the second range in fig. 15 includes only 12 percent instead of the aimed-at 99.73

[20] At this point the reader may wish to consult a simple example worked out on p. 141 of Deming and Birge's *Statistical Theory of Errors*.

percent. Furthermore, most of his ranges would be off center owing to fluctuations in $\bar{X}$. Even on the average, the ranges thus set up would not include 99.73 percent, but something less. For example, the observed average for 1000 such ranges for as many samples of four was in one experiment found to be 93 percent. Of course, it is theoretically possible to choose a coefficient for σ that will overlap 99.73 percent of the chips, on the average, but the errors of the separate ranges would have a larger average than those observed above.

There are many details of interest that might be considered, but for our present purpose it is sufficient to note that the varying experimental ranges have a tendency to hug closer to the ideal limits the larger the sample size used in computing the limits. This fact is of great practical importance, because it shows that if we wish to reduce the chance of making an error in estimating the probability associated with chosen tolerance limits, there is no royal small-sample road for doing this. Even under the simple conditions here assumed, we can improve our estimate *only* by increasing the sample size n. And even with the normal bowl universe one would not likely be satisfied with a sample of less than 1000 and would most certainly require 100 or more if he were trying to set tolerance limits that would insure efficient use of engineering materials. That is to say, even if the properties of materials and manufactured products were in a state of statistical control to begin with, it would still be necessary, in order to acquire the "certain and intimate knowledge"[21] required for setting the most efficient tolerances, to have a sample of at least 100 and more likely a sample of 1000 or more.

Small samples?

It should also be noted that there is no way to form an opinion concerning the errors that might be made in adopting an estimated tolerance range of the form $\bar{X} \pm k\sigma$ *unless we know the sample size n from which it was computed.*

In chapter III we shall be concerned with ranges again, but from the standpoint of the presentation of data.

How Establish Tolerance Limits in the Practical Case?

The necessity for control. Thus far we have considered the method of establishing tolerance limits, assuming that the world is a bowl of chips. Under such conditions, we can increase our knowledge upon which to base tolerance limits only through the process of taking more data, that is, by increasing the sample size. This problem is purely statistical in the sense that any sample of n observed values may be considered as a sample of an indefinitely long sequence of numbers satisfying the requirement that they

[21] Compare this with the quotation at the beginning of this chapter, p. 50.

come from a statistical state of control. Schematically the situation is this:

$$\begin{array}{ccc} X_1, X_2, \cdots, X_i, \cdots, X_n, & | & X_{n+1}, \cdots, X_{n+i}, \cdots \\ \text{Sample} & | & \text{Universe} \\ \text{Past} & \text{Present} & \text{Future} \end{array} \tag{8}$$

How to set tolerance limits L_1 and L_2 upon the basis of the sample, and how to determine the errors that may be expected for samples of n, are problems to the solution of which the mathematical statistician can contribute more than anyone else, *provided,* of course, that the physical state of statistical control represented by drawings from the bowl can be characterized by the mathematics of distribution theory. In fact, in such a state of statistical control, there is, in general,[22] nothing useful that an experimental scientist can tell a statistician about how the n numbers arose beyond the statement that they were drawn from a bowl. Thus we see that since the state of statistical control represents the limit to which one can hope to go in attaining uniformity of quality of product, the setting of the most efficient tolerances reduces in the end to a purely statistical problem.

The statistician does not dare to take it for granted that control exists

Now let us ask: how often in the practical field is one justified in concluding upon the basis of a *small* sample of data that the conditions have been maintained essentially the same in the sense that one would be justified in making predictions as though the sample had been drawn from a bowl? A mathematician obviously can not answer this question; we must appeal to experience for an answer, but in analyzing and interpreting the experience the statistician and scientist must cooperate.

To make our problem specific let us assume that we are given a set of sixteen measurements (table 3) of a physical quantity and that we wish to set tolerance limits for such measurements.[23] What should be our first step?

TABLE 3

6.683	6.681	6.676	6.678	6.679	6.672	6.661	6.661
6.667	6.667	6.664	6.678	6.671	6.675	6.672	6.674

Shall we call in a statistician to proceed as if the sample had been drawn from a bowl, or shall we first call in the scientist who took the measurements to tell us something about them? If we call the scientist, what shall we ask him to do?

[22] The qualifying phrase "in general" is used here to remind us once again that *strictly speaking* we can never be *sure* that we are carrying out any specified measurement or physical operation, including the operation of drawing from a bowl and defined in this monograph as a random one.

[23] It should be kept in mind that we have chosen here to set constant tolerance limits on these measurements, instead of the fluctuating limits given by Student's theory. See the editor's comment beginning on page 59.

Some engineers, scientists, and statisticians make a distinction between the observations of the highly skilled and technically trained research worker and those that an engineer must often work with. They tend to place a kind of halo around the data of *research* as though such data represent the limiting condition that one may hope to attain in removing causes of variability. As noted in chapter I, when scientists think they have done an excellent job measuring some physical constant or property, they have the habit of saying that all the measurements were made under the "same essential conditions." The statistician as a rule not knowing any too much science and the scientist not knowing any too much statistics, the two have often gotten together and agreed, as it were, that the phrase "same essential conditions" can be taken as a password between the two groups. Hence one might conclude that it would be sufficient to ask the scientist if the data of table 3 had been taken under the same essential conditions. If the scientist answers yes, then one might be tempted to turn the problem over to the statistician for him to tell us what he can upon the *assumption* that the 16 data constituted a sample from a bowl of chips.

Engineering and "research" data are not to be regarded differently with respect to the assumption of statistical control. Those who would agree to use this procedure for research data would likely not agree to its use in

TABLE 4

Source	Maximum	Minimum	Average
	Tensile Strength lb. per sq. in.		
No. 1	59 000	45 000	54 000
No. 2	58 500	53 000	56 250
No. 3	56 880	50 000	52 460
No. 4	55 850	47 850	52 890
No. 5	62 140	54 400	57 920
No. 6	62 860	52 150	56 350
No. 7	56 000	50 000	53 000
No. 8	58 000	50 000	55 000
No. 9	61 300	49 000	55 000
No. 10	59 800	50 000	53 970
No. 11	60 000	46 600	52 670
No. 12	58 000	50 000	53 000
No. 13	62 000	51 000	53 000
No. 14	56 640	45 500	51 170
No. 15	61 500	45 000	53 710
No. 16	58 000	50 500	55 500
No. 17	56 160	50 480	52 830
	Average tensile strength		54 040

setting tolerance limits upon the basis of engineering data such as the 5000 observed values of tensile strength of malleable iron castings, table 1, page 54. They would likely question the justification of assuming that these data arose under a state of statistical control. Would one be justified in questioning this assumption? The answer is yes. For example, the reference from which the 5000 data of able 1 were taken gives also the means, maxima, and minima for large samples of similar tests on other material from the same seventeen different sources, as shown in table 4. The total number of tests summarized in table 4 is more than 20,000. Even though the difference between the averages—54,040 lb. per sq. in. and 54,030 lb. per sq. in.—for the data of tables 4 and 1 respectively is not great, I think that both statisticians and engineers would agree that it is pretty likely that the chance cause system behind the 5000 test values was not free from assignable causes, the reason being that the data of table 4 reveal differences that are statistically significant; and since the 5000 data of table 1 came from the same sources, we may perhaps conclude that one is not justified in assuming that they arose under a state of statistical control.[24] This failure to satisfy the criteria of control is a typical characteristic of engineering data.

And now how about the data of "research"? Let us look at some of the series of data taken in pure science to see if they behave as if they had been drawn from a bowl. Let us look at the scientists' measurements of three of

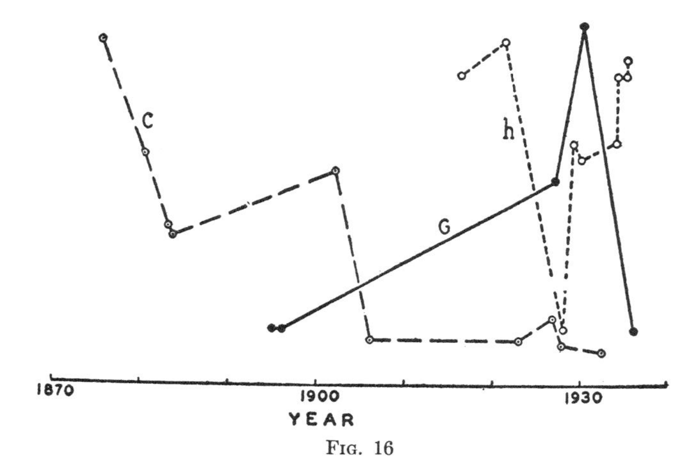

FIG. 16

the seven fundamental constants of physical science, namely the velocity of light c, the gravitational constant G, and Planck's constant h. Certainly such observations are among the elite of all physical measurements. Fig. 16

[24] Incidentally, this comparison between tables 1 and 4 illustrates the loss of information that is apt to result from pooling several sets of data before they can be accepted as homogeneous. *Editor.*

shows the fluctuation in the accepted values of these constants over the past years.[25] The three ordinate scales are not shown since the object here is simply to indicate in a readily comparable way the variations in each of the three sets of measurements over the period from 1870 to 1936. On the evidence here presented, it might be argued that, for the velocity of light c, the accepted measurement seems to be approaching asymptotically some fixed value. This type of argument has, in fact, been advanced by Bavink [26] as indicating the more or less ordered way in which we approach perfect knowledge in physics. The other two curves, however, constitute quite a

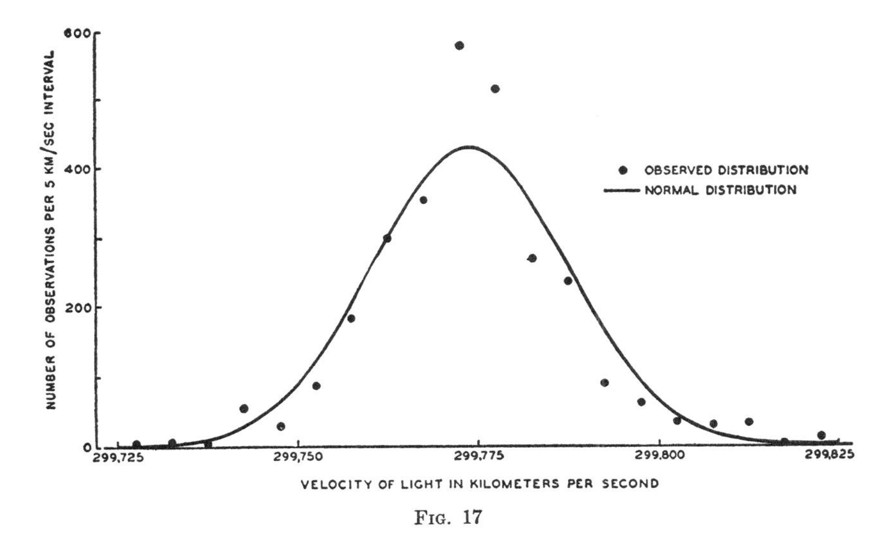

FIG. 17

contrast. Each ends at approximately the level where it began. Physicists pretty generally agree that for each of the three constants, the observed range of variation is so great as to be indicative of "constant" errors.

Moreover it should be kept in mind that the points shown in fig. 16 are averages. Now if we examine the way the single observations are distributed around some of these points, we find further evidence for believing that there

[25] R. T. Birge, "The velocity of light," *Nature*, vol. 134, page 771, 1934. Sten von Friesen, "On the values of fundamental atomic constants," *Proc. Royal Soc. London*, vol. A160, pp. 424–440, 1937. The values of G for 1895 and 1896 are taken from the article "Gravitation" in the eleventh edition of the *Encyclopedia Britannica*. These are the values which the author of the article, J. H. Poynting, thought most likely to be correct at that time (1910). The 1927 and 1930 values are those given in the *Smithsonian Physical Tables*, 1933, while the 1936 value is obtained from "Fundamental physical constants," by W. N. Bond, *Phil. Mag.*, ser. 7, vol. xxii, pp. 624–632, 1936.

[26] Bernhard Bavink, *The Anatomy of Science* (G. Bell and Sons, Ltd., London, 1932).

are assignable causes of variability present. For example, let us consider the last determination of the velocity of light shown on the chart.[27] The total number of repetitive observations in this one point is large—2885 in fact. If these readings could reasonably be treated as though they were a random sample from a normal bowl of chips with an average equal to the true velocity of light, we could be pretty sure that 99.7 percent of a large set of future observations by this method would fall within the range $\bar{X} \pm 3\sigma$. But as is almost always true when a large sample is available, these 2885 observations do not give much evidence of having come from a normal universe. Fig. 17 compares the observed distribution of these observations with the fitted normal curve. The χ^2 test tells us that the probability of getting a deviation from normality (as measured by χ^2) as large as or larger than that observed, is too small to be read from the tables of χ^2. Hence if one wished to set up valid tolerance limits on future observations of the velocity of light, he would be unwise to use a rule based upon the assumption of normality.

But—and this is the most important question—are we justified in believing that these data constitute a random sample from *any* universe, normal or otherwise? Dare we assume that they arose from a constant system of chance causes of variation or, in other words, from a state of

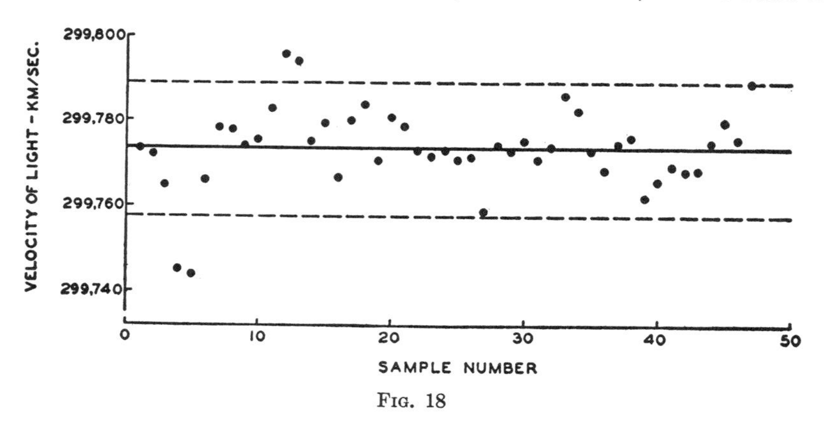

FIG. 18

statistical control? Suppose we let the data speak for themselves when successive groups are plotted in the form of a control chart,[28] fig. 18. The chance of one of these averages going outside the dotted limits if the samples

[27] Michelson, Pease, and Pearson, "Measurement of the velocity of light in a partial vacuum," *Astrophysical Journal*, vol. 82, pp. 26–61, 1935.

[28] Criterion I as described on p. 309 of my *Economic Control of Quality of Manufactured Product* is here used (cited on p. 23).

had come from a constant chance cause system (even though we do not know the distribution function for the cause system, i.e., even if we do not know the form of the universe) is not much different from .003. Four points outside the limits, in a total of forty-six, is not a very likely event on the assumption of a constant system of variation. What is the practical significance of this fact for our present story? It is simply this: *Whereas there is safety in numbers when setting tolerance limits on the basis of a sample from a bowl, that same degree of safety does not exist when the samples are not so drawn.* My own experience has been that when data behave as they do in fig. 18 it never pays to rely upon numbers alone.

Now let us look at another point in fig. 16, this time the maximum point shown on the G curve. This value, 6.670×10^{-8} cm^3 g^{-1} sec^{-2}, is that given by Heyl.[29] It was derived from the three sets of measurements shown

TABLE 5

VALUES OF G IN UNITS OF 10^{-8} CM^3 G^{-1} SEC^{-2} (HEYL[29])

Gold	*Platinum*	*Glass*
6.683	6.661	6.678
6.681	6.661	6.671
6.676	6.667	6.675
6.678	6.667	6.672
6.679	6.664	6.674
6.672		

in table 5 corresponding to experiments using platinum, gold, and glass spheres. The value given by Heyl is obtained by weighting the data for

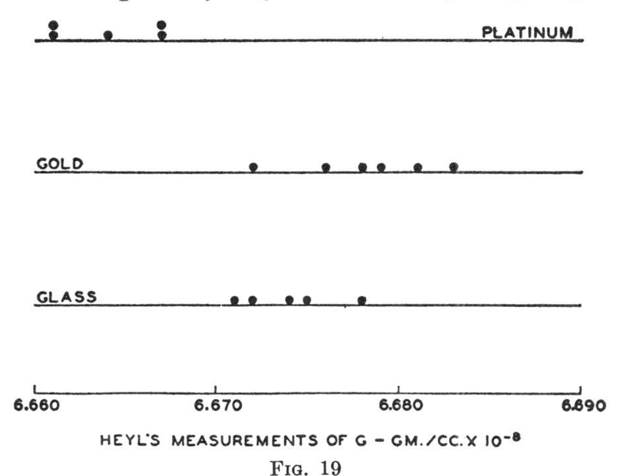

FIG. 19

the gold spheres by one third and the other data by unity. The data of table 5 are shown graphically in fig. 19. Certainly we need no refined

[29] Paul R. Heyl, *Bureau of Standards Journal of Research*, vol. 5, pp. 1243–1290, 1930.

statistical tests to convince us that such a set of sixteen data is very unlikely upon the assumption that a constant system of chance causes is the source of the observed variability. Heyl says: "The different results obtained with the various materials used for the small masses are yet to be explained, but evidence is given that this difference is not to be ascribed to the nature of the material." The point I wish to stress is that here again we have a sample of measurements among the most elite of pure science that do not seem to behave like drawings from a bowl of chips.

Where does the statistician's work begin? Now let us return to the question, given the problem of setting a tolerance range on measurements of the type shown in table 3, shall we first call in a statistician or shall we first call in the scientist who is an authority in the field from which the data came? The evidence (fig. 19) of lack of control in the measurements of the constants c, G, and h may well serve to shake our faith in a scientist's judgment that the conditions have been maintained essentially the same, and in regarding his statement as a satisfactory basis for turning the data over to a statistician to be treated as if it were a sample from a bowl. As the reader may have already noted, the sixteen measurements of table 3 (p. 64) are the same as those of table 5, except for a constant multiplier; hence, after our experience with fig. 19, we are able to say something concerning the question that was asked regarding the numbers in table 3; it would be hazardous to consider them as a sample drawn from a normal bowl or any other kind of bowl, no matter who took the data.

The statistician is supreme after statistical control has been established; until then the scientist and the statistician must cooperate

In the light of such experience in the investigation of available measurements of the physical constants and in the light of my experience in the study of samples of measurements of quality in engineering, I feel that before one turns over any sample of data to the statistician for the purpose of setting tolerances he should first ask the scientist (or engineer) to cooperate with the statistician in examining the available evidence of statistical control. The statistician's work solely as a statistician begins after the scientist has satisfied himself through the application of control criteria that the sample has arisen under statistically controlled conditions. The case is something like the old story of Pat, the Irishman, who had been in this country only a few months and in the meantime had located a job as a hod-carrier when his friend Mike arrived. "Pat," says Mike, "and what are you doing?" To which Pat answered, "Sure an' I have an easy job. I carry the bricks up four flights of stairs and the man up there does all the work." In much the same sense the scientist must carry his data through several control criteria before handing them over to the statistician to use in setting tolerances.

There still remains the question how we are to set tolerance limits when the chance cause system is not in a state of control. Certainly the engineer and the scientist both must set tolerance ranges within which measurements of physical constants and properties may be expected to lie even when conditions indicate that the state of statistical control has not been attained. Perhaps enough has been said to show that the establishment of tolerance limits under conditions that are not statistically controlled is not a problem to be turned over to the statistician to solve by himself on the assumption that the available data can be treated as a sample from a bowl.

Further Considerations Regarding Tolerance Limits

As a starting point for what follows, we need to look more critically than heretofore at the requirements that tolerance limits must meet in the process of mass production of interchangeable parts. So far, we have spoken only of tolerances expressed in terms of the *measurements* of some quality characteristic. It has been tacitly assumed that if the *measurements* of a quality characteristic on two or more pieces of a given kind of product fall within their tolerance limits, then the quality of all of these pieces of product falls within these same tolerance limits. Obviously, however, this assumption may not be justified because the measurements may be, as we say, "in error." Hence we need to take into account the difference between the customarily accepted concept of the true value X' of a physical quality and a measurement X of this true value.

For example, if we have two pieces of product O_1 and O_2 of the same kind, we customarily assume that the values X'_1 and X'_2 of their true quality characteristics must both lie within some tolerance range

$$X' = L_1 \quad \text{to} \quad X' = L_2 \tag{9}$$

in order that the objects be interchangeable in assembly and use in respect to the quality characteristic X'. Likewise the desired physical state of statistical control is assumed to be expressible in terms of a sequence of numbers representing true values of the quality characteristic X' for a sequence of objects:

$$X'_1, X'_2, \quad \cdots, X'_i, \quad \cdots, X'_n, \quad X'_{n+1}, \quad \cdots, X'_{n+i}, \quad \cdots \tag{10}$$

These are formal expressions of the fundamental requirements for economic mass production of interchangeable parts.

Let us now look a little closer at this concept of a true value X' as here used. How is one to determine whether the true value of the quality characteristic lies within a given range? If one can not discover the true value, then of what practical use is the concept of true value? In answer we shall see that the concept of true value leads us to *choose* operationally verifiable criteria that measurements of a quality characteristic must satisfy in order that they may be considered to be measurements of the true value X'. These criteria, as we shall see, include those for control of any method of measurement and those for checking the consistency between measurements by different methods.

The concept of the true value leads to operationally verifiable criteria regarding measurements

To begin, let us note that corresponding to every concept of a true measurable quality characteristic X', such for example as length, there are usually several assumed methods of measurement. For example, a method may involve the use of (*a*) an ordinary rule, (*b*) a micrometer, (*c*) a traveling microscope, or (*d*) triangulation. Presumably the operation of measurement by each method can be repeated again and again at will so that corresponding to any true value X' there are potentially as many infinite sequences of measurements as there are assumed methods of measuring. Schematically the situation is this:

$$X' \overset{O}{\rightarrow} \begin{cases} X_{11},\ X_{12},\ \cdots,\ X_{1i},\ \cdots,\ X_{1n},\ X_{1,\,n+1},\ \cdots,\ X_{1,\,n+i},\ \cdots \\ X_{21},\ X_{22},\ \cdots,\ X_{2i},\ \cdots,\ X_{2n},\ X_{2,\,n+1},\ \cdots,\ X_{2,\,n+i},\ \cdots \\ \vdots \\ X_{i1},\ X_{i2},\ \cdots,\ X_{ii},\ \cdots,\ X_{in},\ X_{i,\,n+1},\ \cdots,\ X_{i,\,n+i},\ \cdots \\ \vdots \end{cases} \tag{11}$$

where the symbol $\overset{O}{\rightarrow}$ stands for an operational meaning of measuring X'.[30] The diagram of sequences (11) thus portrays the fact that each method of measuring gives rise to a sequence of observations, and if there are several methods, there are as many sequences. However, in order for such a set of sequences (11) to constitute *the* operational meaning of measuring the true value X', each sequence must represent a statistically controlled condition, and the statistical limits of the averages of the first n terms of these sequences

[30] The transition from the sequences (11) to the equalities (12) constitutes a bridge between the abstract concept of X' and a physically verifiable operation of measurement. In chapter IV we shall see how one can pass from physical to practical verifiability.

There is no way of giving practical verification to eqs. (12), yet these equalities are physically verifiable in the sense that to any series of measurements in the sequences (11), one more observation can always be added, and the average taken. But the average $\bar{X}$ of any finite number of measurements is not the same thing as the theoretical limit $\bar{X}'$, even if for some value of n, $\bar{X}$ happens to be equal numerically to $\bar{X}'$. This brings up the question, when can one say that eqs. (12) are true? For an answer one must introduce the concept of *practical verifiability*, involving certain limitations expressed in terms of tolerance ranges. This matter will be considered carefully in chapter IV. *Editor.*

as n approaches infinity must be equal; or expressed formally,

$$\bar{X}'_1 = \bar{X}'_2 = \cdots = \bar{X}'_i = \cdots \qquad (12)$$

Standard methods of measuring. In practice it is customary to choose one of the methods of measurement as a standard. For this method we may write the potentially infinite sequence of measurements of the true value X' as

$$S_1, S_2, \cdots, S_n, S_{n+1}, S_{n+2}, \cdots \qquad (13)$$

to set if off from the others. Theoretically, this sequence in order to serve as a basis for comparison should be *random in the sense that it is representative of a state of statistical control.* Requirement (12) then reduces to

$$\bar{X}'_i = \bar{S}' \qquad (i = 1, 2, \cdots) \qquad (14)$$

As statisticians we might have introduced the requirement in (12) and (14) that these statistical limits be equal to the true value X', or as we sometimes say, that the method of measurement shall not be biased. Operationally, however, we have no physical or experimental way of getting at X' except through measurement, and hence the requirements (12) and (14) are here expressed in terms of measurements alone. It should be noted, of course, that (12) and (14) express the requirements in a formal and hence abstract manner. We shall later consider the practically verifiable operational meaning of these expressions in use (chapter IV).

Let us pause for a moment to examine some of the proposed standard methods of measuring a quality characteristic such as length to see what criteria such measurements must satisfy. These methods of measuring are usually divided into two classes: those using some arbitrarily chosen physical object such as the Imperial Standard Yard and the International Prototype Metre and those using some natural phenomenon such as the wave length of light.

First let us consider the requirement of randomness or statistical constancy of the standard sequence (13) when applied to a typical standard method of measurement. To begin, we shall choose an operationally verifiable criterion for control in the sense considered in chapter I. The layman might expect that, having chosen a criterion of control, it would be quite simple to find a standard experimental sequence that satisfies the one chosen. For example, he might expect Michelson's measurements on the velocity of light to constitute such a sequence for length. A glance, however, at the control chart record (fig. 18, p. 68) for these measurements of the velocity of light should be sufficient ground for believing that this method of measurement, at least as represented by Michelson's data, does not satisfy the criterion for control here chosen. Since we do not find evidence of statistical control in measurements of some of the most important physical constants, it would seem that the first real problem in establishing a *standard* sequence in terms of measurements of a physical phenomenon is to detect and eliminate assignable causes of variation until one can be reasonably sure that he has attained a state of statistical control in the measuring process.

Statistical control the first step in the establishment of a standard sequence

Now let us see what the situation is for measurements in terms of arbitrarily chosen physical standards. Some interesting results have recently been given by J. E. Sears, superintendent of the metrological department of the National Physical Laboratory. In addition to the Imperial Standard Yard, there are in existence at least four Parliamentary copies. Table 6 shows the observed differences in millionths of an inch between the length of the Imperial Standard Yard I and the copies P.C. 2; P.C. 3; P.C. 5; and P.C. VI. Sears places the observations on P.C. 3 in 1876 and those on P.C. 3 and P.C. 5 in 1892 under suspicion, and hence he argues that according to the results shown in this table the lengths of the bars P.C. 2, P.C. 3 and P.C. 5 have remained in close agreement with that of the standard. However, he points out that not only the evidence given in table 6 but also other evidence cited in his article indicates that P.C. VI contracted over this period in an exponential manner so as to approach the asymptotic difference of -228×10^{-6} inch which the bar has now reached. Sears points out that

TABLE 6

Difference in Millionths of an Inch

Comparison	*1852*	*1876*	*1886*	*1892*	*1902*	*1912*	*1922*	*1932*
P.C. 2 —I	+21	+36	—	+ 6	—	− 23	− 19	− 39
P.C. 3 —I	−33	+57	—	+55	—	− 49	− 61	−111
P.C. 5 —I	−55	−33	—	+70	—	− 43	− 23	− 47
P.C. VI—I	—	—	−3	—	−192	−215	−217	−234

the bar P.C. VI was made several years after the others and argues that perhaps the reason why the change in length is noted only in the case of P.C. VI is that the others had reached a stable state before the measurements in table 6 were taken. Of course, another explanation might be that the earlier bars, including the Imperial Standard, have been shrinking at the same rate.

For our present purpose, the point I wish to make is that there is evidence for believing that the use of such arbitrarily chosen physical standards of length can not be expected to give a random test series, at least until the physical standards themselves are several years old. The question of how many years are required in any given case can be determined only through a study of the test results at intervals over this period to determine whether they give evidence of having attained a state of statistical control. The initial measurements obtained by the use of such standards certainly do not give evidence of having arisen under statistically controlled conditions.

What is more important, however, from an operational viewpoint, is to scrutinize the measurements that are obtained by a given method in order to determine not only whether they have arisen under a state of control but also whether they are significantly different from those obtained by other accepted methods; in other words, to determine whether such measurements satisfy the requirement (12) or (14) as the case may be.

The use of any one of the duplicate physical standards of length in table 6 is capable of giving an infinite sequence; hence corresponding to the measurements of a length by the five standard bars, we should have five sequences of the form shown below:

$$X' \overset{O}{\rightarrow} \begin{cases} S_{11}, S_{12}, \cdots, S_{1i}, \cdots, S_{1n}, S_{1,\,n+1}, \cdots, S_{1,\,n+i}, \cdots \\ S_{21}, S_{22}, \cdots, S_{2i}, \cdots, S_{2n}, S_{2,\,n+1}, \cdots, S_{2,\,n+i}, \cdots \\ \vdots \\ S_{i1}, S_{i2}, \cdots, S_{ii}, \cdots, S_{in}, S_{i,\,n+1}, \cdots, S_{i,\,n+i}, \cdots \\ \vdots \end{cases} \tag{15}$$

Presumably, duplicate copies of a standard should be interchangeable in terms of the infinite sequences (15) that characterize them in an operational way. Hence, from the viewpoint of statistical theory, the requirements imposed on the sequences in (11) are different from those imposed on the sequences in (15). It follows that sequences (11) and (15) must both satisfy the conditions (12), and in addition that the sequences in (15) must also satisfy the condition

$$f(S) \equiv f_1(S_1) \equiv f_2(S_2) \equiv \cdots \equiv f_i(S_i) \equiv \cdots \tag{16}$$

which is supposed to symbolize the requirement that the sequences in (15) may all be considered as random sequences from the *same* universe.

Setting tolerance limits when control is lacking. Now we are in a position to consider in what practical sense we can set tolerance limits on the "true value" X' under practical conditions. The first thing we must do is to ascribe an operational meaning to the measurement of the true value X' that satisfies eqs. (12). If, in a practical case, one knew that each of the sequences corresponding to the assumed methods of measuring the true value was random, and that the requirements (12) and (16) were satisfied, then one could proceed in setting tolerance limits as he would for samples drawn from a bowl. Evidence has been given, however, to indicate that requirements (12) and (16) are not met even for the simple case of measuring a length. How then shall we proceed?

All that we can do here is to consider some of the general principles that we must take into account in the establishment of tolerance ranges under conditions of lack of control. To begin with, we must give attention to the meaning, validity, and efficiency of the range.

Meaning. In establishing a tolerance range for drawings from a bowl universe, we attempt to estimate from a sample of n a range that may be expected to include $(1 - p')N$ of N future numbers drawn from the same bowl. In contrast, let us consider the problem of establishing a tolerance range on the tensile strength of malleable iron on the basis of the 20,000 measurements given in table 4, p. 65. The tensile strength of malleable iron is more complicated and much less definite than drawings from a bowl, on two scores. In the first place, one must choose the methods that are to be included in the sequences (11), which define the operation of measuring the quality characteristic of tensile strength. In the second place, one must decide which sources of supply of the material are to be covered by the tolerance. One must, in other words, define the operations of choosing the material to be included. Obviously, establishing a tolerance range for one of the sources of material in table 4 would be quite a different problem from establishing a tolerance range for all of the sources noted in this table, which in turn is a simpler problem than establishing a similar tolerance range applicable to all sources that might be included in the future. We shall give more critical attention to the operational meaning of tolerance ranges in chapter IV. We should note here, however, that in the case of a bowl we may conceivably set a tolerance range that may be expected to include $(1 - p')N$ of future drawings from the same bowl, whereas under non-controlled conditions we can conceive of establishing a tolerance range only in the sense of finding a range such that the *probability* of future observed values falling within this range *can not be less than* $1 - p'$.

Validity. In setting tolerance ranges for future drawings based upon a sample of size n drawn from a bowl, it was pointed out that the tolerance range might involve a huge error if fixed upon the basis of a small sample (fig. 15, p. 62). Under conditions of lack of control the chance for error is even greater. For example, if one were to set a tolerance range for the tensile strength of malleable iron upon the basis of a sample of size n from the source having the smallest range in table 4 (p. 65) it would obviously involve a huge error irrespective of sample size if applied to any of the other sources shown in the table.

In the majority of practical instances, the most difficult job of all is to choose the sample that is to be used as the basis for establishing the tolerance range. If one chooses such a sample without respect to the assignable causes present, it is practically impossible to establish a tolerance range that is not subject to a

Choosing the sample

huge error. Before choosing the sample, therefore, it is desirable to try to detect the presence of assignable causes and to discover the nature of these so that their influence may be foretold. The operation of quality control,[31] as well as tests for significant differences, is of great use in this connection if the tolerance range is to be set so as to include the variability that may arise if none of the assignable causes is removed. Under such conditions, one must try to choose the tolerance limits L_1 and L_2 so that under the worst conditions that one may reasonably look forward to in the light of a study of the nature of the assignable causes present, not more than $p'N$ of any group of N observations are expected to fall outside the limits L_1 and L_2 in a series of N trials.

Thus in setting tolerances for the tensile strength of malleable iron where it is desired to include all of the 17 sources in table 4 under the assumption that they are to remain as uncontrolled as they are, one would simply take into account the best and worst sources as a basis for setting the tolerance limits. Then, since it is likely that each of these two sources is not statistically controlled, one would have to allow for the effects of assignable causes as best he could. Oftentimes under such conditions the maximum and the minimum in the best and worst sources respectively are of more importance than any other statistics of these distributions for indicating the range in which most of the future observations will lie.

Emphasis should be placed upon the fact that in the use of statistical tests for significant differences it is necessary to use large enough samples to reduce to a satisfactory level the risks of making errors in judgment. The reason for such action is similar to that for going to a sample size between 100 and 1000 in trying to establish a tolerance range even in the simplest case of drawing from a normal bowl, as was pointed out in the discussion of fig. 15, p. 62. Also I think it is important to note how extensive the series of measurements apparently must be before we can hope to gain much by trying to analyze a set of data as though it were a sample from a bowl. For example, in the beginning of any investigation involving the measurement of a "true" value there are usually only a few known methods of measuring the quantity in question. At least in the field of physical and chemical science, the requirement of *consistency* [32] between the results obtained by

[31] See, for example, H. F. Dodge, "Statistical methods and specification of quality," *Bulletin of the American Society for Testing Materials*, No. 85, pp. 17–21, 1937.

[32] This term as here used means agreement or harmony of the sequences among themselves as parts of the assumed operational meaning of measuring the true value. A chosen set of sequences is assumed to be consistent with respect to any specified statistic of the sequences when the observed differences in the values of this statistic calculated from the observed portions of the sequences are not greater than may reasonably be left to chance as determined by some chosen criterion.

different methods has been a powerful influence in directing attention to the so-called constant errors. It would appear that, in general, it is of little value to make very large numbers of measurements by any one method until it has been found to give results that are more or less consistent with those obtained by other methods. If, however, a large number of measurements are to be made as, for example, in the measurement of the velocity of light, it would seem that much is to be gained by applying statistical criteria of control for detecting assignable causes of variability, because in no other way apparently can we reach the state of statistical control and maximum validity in prediction.

Consistency between different methods of measurement is of great importance

Efficiency. Under conditions that are not statistically controlled, the tolerance limits must be set much farther apart than would be necessary if the operation of statistical control were applied to detect and weed out unnecessary causes of variability. Setting an unnecessarily broad tolerance range naturally leads to an inefficient use of materials. For example, in the design of ships,[33] or structures of any kind, if the engineer makes the tolerances unnecessarily wide, such action results in the use of more material than is necessary. It should, of course, be noted that efficiency in the sense here used is limited to the concept of minimizing the quantity of material used and hence is to be differentiated from the broader concept of economic use which must take into account efficient use of material as only one of several factors.

If we are going to make the most efficient use of material, we must close up on the tolerances as far as it is economical to go. In this process, we must make use of two kinds of statistical criteria: (*a*) those involved in the operation of control, and (*b*) those required to test the consistency between the sequences used in giving operationally definite meaning to the true value X' schematically illustrated in (11) and (15), pp. 72 and 75. Criteria under (*b*) are obviously those for testing significant differences in averages and for testing whether it is reasonable to believe that a given set of sequences came from the same state of statistical control. This progress toward the ultimate goal of efficient use of raw materials through reduction of tolerances to an economic minimum necessarily involves extensive use of tests for significant differences.

Someone may ask, why go further than scientists have gone in trying to attain random sequences of physical measurements satisfying the criteria (12) and (16)? The answer is that, in just the same way that industrial applications of scientific principles have brought more and more stringent require-

Stringent requirements in industry

[33] *Cf.* W. P. Roop, "Features of practice affecting design," a paper read at the annual meeting of the Society of Naval Architects and Marine Engineers, 1936.

ments on accuracy in measurement, so it is that any further steps toward attaining maximum efficiency in the use of materials will bring additional requirements on the methods of measurement in regard to the state of statistical control and maximum consistency, both of which will necessitate the extensive use of statistical theory and technique.

From what has been said in this chapter, it appears that we must gain a much more intimate knowledge of the properties of materials than we now have if the engineer of the future is to minimize tolerance ranges and thereby attain maximum efficiency in the use of materials. Furthermore it must be apparent that this ideal can be attained only by the application of statistical theory in establishing criteria for control and other criteria for testing consistency between methods of measurement. Even in establishing tolerances under conditions that are not statistically controlled, it is to the engineer's advantage to use statistical technique as an aid in segregating assignable causes of variability; and when a state of statistical control is reached, the setting of tolerance limits becomes a purely statistical problem.

CHAPTER III

THE PRESENTATION OF THE RESULTS OF MEASUREMENTS OF PHYSICAL PROPERTIES AND CONSTANTS

A Worthy Goal:

"When you can measure what you are speaking about and express it in numbers, you know something about it, but when you cannot measure it, when you cannot express it in numbers, your knowledge is of a meagre and unsatisfactory kind."

LORD KELVIN

But:

" . . . knowing begins and ends in experience; but it does not end in the experience in which it begins."[1]

C. I. LEWIS, *Harvard University*

THE NATURE OF THE PROBLEM

Increased knowledge of quality necessary. To make the most efficient use of both raw and fabricated materials, the engineer needs to increase his present knowledge of their quality characteristics. In fact, he must know more in the future than anyone now knows about the variability of almost every such quality characteristic. Needless to say, the measurements of physical properties and constants made by the scientist and engineer in the research laboratory contribute materially to such knowledge. However, as we have seen in the previous chapter, the engineer must also have more knowledge than he now usually has about the variability of each quality characteristic of his product under *commercial conditions of production* if he is to be able to set the most economic tolerance limits on each characteristic. The object of this chapter is to consider how an understanding of statistical theory may help one to present the results of measurement in a way that will contribute most effectively to the *knowledge* that the engineer must have if he is to establish tolerances that will make possible the most efficient use of his materials. The emphasis throughout this chapter is accordingly to be placed upon the presentation of observed results as an *evidential basis for knowledge;* in fact, the title might well have been, "The Presentation of Data as Evidence."

Some considerations of summaries of the density of iron. As an example, let us assume that we wish to make use of pure iron in such a way that its density is one of the quality characteristics upon which we wish to set economic tolerance limits. Since this is a property of pure iron that has

[1] "Experience and meaning," *The Philosophical Review*, vol. xliii, p. 134, 1934.

been studied at length by different scientists and engineers, we may expect to find the results of their work summarized in standard tables of physical constants. Let us see if such summaries provide adequate knowledge for establishing economic tolerance limits. If one looks in the Smithsonian tables, for example, he finds that the density of pure iron at ordinary temperature [2] is given as

$$7.86 \text{ gm/cm}^3.$$

Obviously this single value does not provide a basis for setting tolerance limits because it does not indicate how much variability may be expected. If one looks in another recent and authoritative table,[3] he finds that the density of pure iron at approximately room temperature is given as

$$(7.871 \pm 0.002) \text{ gm/cm}^3.$$

Does such a summary provide an adequate basis for establishing a tolerance range? Let us assume as a basis for our discussion of this question that 7.871 is an estimate of the true value of the density and that 0.002 is an estimate of the probable error. Suppose now that we want to set a 99.7 percent tolerance range. Does the information that we have found provide an adequate basis for establishing such a range?

Size of sample must be stated even in controlled experiments. $X \pm \Delta X$ not enough

In the light of the experimental results presented in fig. 15 (p. 62) it is apparent that the error that might be made in setting a tolerance range upon the basis of such evidence, even though the original data were normally distributed without constant error about the true value, may be quite large unless the estimates of the expected value and probable error are based upon a large sample. In other words, we see that *even under idealized conditions* of sampling from a normal bowl universe, it is necessary for one to know the size of the sample if he is to form a reliable estimate of the maximum error that may be expected in the estimated tolerance range. Hence it appears that a summary of the measurements of a quality characteristic X either in the form X or $X \pm \Delta X$ does not in itself provide a satisfactory basis for setting tolerance ranges even though it be known that the quality characteristic X is in a normal state of statistical control.

There is, however, a much more important reason why such presentations of data are inadequate; as emphasized in the previous chapters, measurements of physical properties and quality characteristics, including some of the most refined physical measurements, are not ordinarily in a

[2] *Smithsonian Physical Tables*, 8th revised edition (The Smithsonian Institution, Washington, 1933), p. 160.

[3] *Physical Constants of Pure Metals*, The National Physical Laboratory (His Majesty's Stationery Office, London, W.C. 1, 1936), p. 6.

state of statistical control. As a further illustration of this fact, let us look at fig. 20 which shows the ranges for several different determinations of the velocity of light as given in a recent article.[4] The length of the vertical line in each case is proportional to the recorded range. If we compare the succession of ranges in fig. 20 with those shown in fig. 14 (p. 59), neglecting for the moment the fact that the two sets of ranges have not been calculated in the same way, I think it is obvious that the ranges in fig. 20 do not appear to behave on the whole like those in fig. 14; in

A more important defect in summaries in the form $X \pm \Delta X$

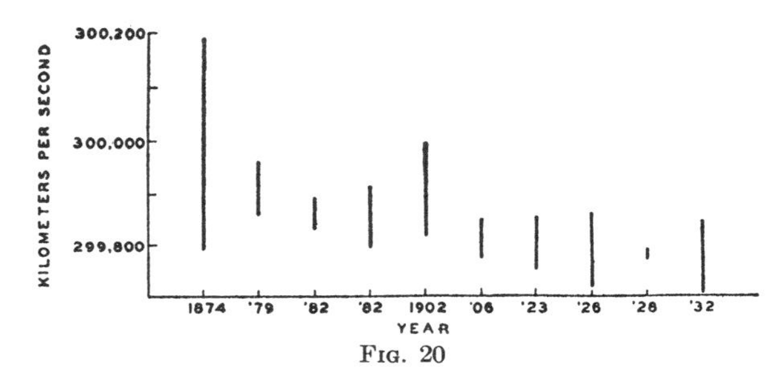

FIG. 20

particular, the succession of ranges in fig. 20 does not appear to center about some constant expected value. If we were to construct corresponding sets of ranges for the measurements of the gravitational constant G and Planck's constant h (fig. 16, p. 66), we should find that they also do not behave like the normal bowl ranges in fig. 14. This illustrates the simple fact that the meaning or interpretation of any summary in the form $X \pm \Delta X$ depends upon whether the original data arose under a state of statistical control; such a summary of data does not provide an adequate basis for setting an efficient tolerance range.

Still other arguments against $X \pm \Delta X$

There are, however, certain other reasons why the customary practice of summarizing data in the form $X \pm \Delta X$ does not provide the necessary basis for setting economic tolerances. Perhaps the chief among these is the fact that different methods are often used in making a summary of a given set of data in the form $X \pm \Delta X$. In other words, starting with an original set of data, different scientists may use different methods in estimating the true or expected value, and the probable error or some other measure of dispersion. They may also choose a probability other than $\frac{1}{2}$ in arriving at ΔX. Hence $X \pm \Delta X$ as used in practice does not always have the same meaning.

[4] Sten von Friesen, "On the values of fundamental physical constants," *Proc. Roy. Soc. London*, vol. A160, pp. 424–440, 1937. Only the first five ranges in fig. 20 are taken from von Friesen's article. The other ranges take into account different data. There are two sets of measurements for the year 1882.

This immediately suggests a question that has been the subject of extensive investigation in the field of statistics—what method should be used in estimating?

Enough perhaps has been said to indicate the nature of the problem that one encounters in trying to make efficient use of data as customarily summarized. Owing to the fact that such summaries do not usually provide adequate knowledge, it is necessary for engineers to consider the fundamental problem of how one should tabulate data on the quality characteristics of raw and fabricated materials so as to provide an evidential basis for the maximum amount of knowledge that one wishes to convey for the establishment of valid and efficient tolerance limits.

The importance of the problem of presenting data. Before we plunge into a discussion of the technical aspects of the problem of presenting data, it is fitting that we mention briefly some of the other ways in which this problem has come to the attention of engineers. In December 1926 a Sectional Committee on Standards for Graphical Presentation was organized under the procedure of the American Engineering Standards Committee. The scope of this committee's work included the development of the basic principles that should be used in the preparation of scientific and engineering graphs. One of the problems early brought to the attention of this committee was that of presenting data graphically in a way to provide as much knowledge as possible about the variability of measurements. Fig. 20 serves as a simple illustration of an attempt of one scientist to provide graphically some knowledge about the variability of the measurements of the velocity of light. It is obvious, however, that we must find a satisfactory method of summarizing data analytically along the lines called for in the previous section before we can summarize the results graphically. Hence it is reasonable to say that the satisfactory solution of the problem of determining how best to present experimental data graphically must await a satisfactory solution of the problem of presenting the numerical aspects of data so as to make a maximum contribution to knowledge in the sense discussed in this chapter.

About 1930 the American Society of Mechanical Engineers and the American Society for Testing Materials jointly sponsored the formation of a cooperative [5] committee to consider, among other things, the problem of applying statistical theory in the presentation of the great quantities of data taken by engineering and scientific groups in the study of the physical properties of raw and fabricated materials, such data being intended for

[5] This committee on the application of statistics in engineering and manufacturing is now jointly sponsored by the two engineering societies named above and by the American Statistical Society, the American Mathematical Society, and the Institute of Mathematical Statistics. The American Society for Testing Materials also organized a committee about this time on the interpretation and presentation of data to consider the specific problems arising in their society.

later use in the establishment of economic tolerances. A vast amount of data of this nature is summarized each year, and it is therefore of considerable commercial importance to find the most useful method of doing the work. Having been closely associated with these committees since their organization, I shall try to present in this chapter some of the basic principles that have been found useful in guiding the choice of a method for presenting the kinds of data ordinarily obtained by these committees.[6]

We should note at the beginning that the consideration to be given here is limited in a fundamental way; it has to do with the presentation of data only from the viewpoint of providing knowledge and not from the viewpoint of securing an emotive reaction on the part of the one who reads the results after they are presented. To make clear the significance of this limitation it is desirable to consider briefly the way in which a summary of a group of data is a kind of language, and hence must comply with the accepted requirements of a meaningful language if it is to be scientific.

The presentation of data from the viewpoint of language. There is scientific language and emotive language. A presentation of the result of measurement by an author serves as a language of communication between the author and his readers. Now there are at least two distinct uses of such language:[7] (*a*) to communicate information or knowledge; and (*b*) to arouse an emotional attitude in a reader or to influence his action in any way other than by the information transmitted. These two uses of language have been referred to as the *scientific* and *emotive* respectively.

The statistician's language is sometimes emotive

In the presentation of scientific and engineering data for the purpose of providing the reader with a knowledge of engineering materials, it is necessary that the language used be scientific and not emotive. A statistician must keep this requirement in mind if and when he steps in to help the scientist and engineer. Strange as it may seem in the face of this situation, the statistician sometimes rushes in to help the scientist and engineer do a scientific job and forgets that a lot of his professional lingo is more emotive than scientific: witness, for example, the statistician's use of such phrases as statistical facts, confidence limit, probable error, most probable value, and best estimate, to mention only a few. Some of these terms—particularly most probable value and probable error—have been taken over quite extensively by the scientists. Sometimes scientists even add new terms of their own, such as "even-bet" error instead of probable error.

[6] Such summaries are presented not only in original memoirs and reports but also in engineering handbooks and tables such as the *Smithsonian Physical Tables*, the *International Critical Tables*, etc.

[7] See for example L. S. Stebbing, *A Modern Introduction to Logic* (Thomas Y. Crowell Co., New York, 1930), pp. 10–21.

To illustrate the difference between some of the emotive terms used by the statistician and some of the terms in current use by physical scientists, let us consider what the scientist sometimes says about the measurement of a physical constant such as the velocity of light. In the first edition (1926) of the *International Critical Tables*, we find tabulated what the editors of that publication call the accepted, conventional, or defined values of the physical constants to be regarded as exactly correct for purposes of computation. In what respect does such an accepted value differ from a most probable value or from a best estimate? If a reader, ignorant of the technical meaning of best, finds two tables, one giving the best estimates and the other the accepted values, is he not likely to *feel* that the best estimates should be better than simply accepted values, and make his choice accordingly? Certainly the use of the term "best" introduces a large emotive element not present in the term "accepted value," but this emotive element does not contribute to scientific knowledge in that it does not have an operationally definite meaning in terms of future experience. Scientific statements presumably state something about an object or physical phenomenon that can be tested experimentally by an observer and thus can be shown to be either true or false. Hence, in the use of statistical techniques in the presentation of scientific data, we must be careful to give scientific meaning to all terms used. Although meaning is thus an essential component of knowledge, it is not the only one—in fact there are two others; and we must thoroughly understand all three and their interrelations in order to consider intelligently the role that statistical theory may be made to play in the presentation of scientific data.

Three Components of Knowledge—Evidence, Prediction, Degree of Belief

Basic to all that follows is the concept of knowledge here adopted. In line with the statement quoted from C. I. Lewis at the beginning of this chapter, I shall assume that knowledge begins and ends in experimental data but that it does not end in the data in which it begins. From this viewpoint, there are three important components of knowledge: (*a*) the data of experience in which the process of knowing *begins*, (*b*) the prediction P in terms of data that one would expect to get if he were to perform certain experiments in the *future*, and (*c*) the degree of belief p_b in the prediction P based on the original data or some summary thereof as evidence E. These three components are schematically illustrated in fig. 21. Knowledge begins in the original data and ends in the data predicted, these future data constituting the operationally verifiable meaning of the original data. Since, however, inferences or predictions based upon experimental data can never be certain, the knowledge based upon the original data can inhere in these data only to the extent of some degree of rational belief. This follows from Postulate II (p. 42), in which it is assumed that there is an objective degree of rational belief p'_b belonging to the relation between any prediction and the original data upon which the prediction is based.

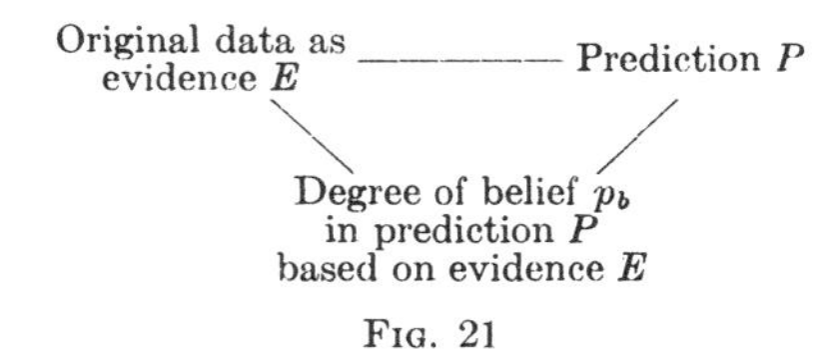

FIG. 21

What has just been said about the three components of knowledge may appear to the practical engineer or statistician as being abstract and somewhat formal until he considers how they are met in everyday experience. For example, I might say, "It is going to rain day after tomorrow." That statement has a definite predictive meaning in the sense that you can test it in the future. However, it doesn't convey much knowledge, since I have no standing as a weather prophet. You may therefore ask what makes me think that it is going to rain day after tomorrow. That is, you *ask for my evidence.* Given the evidence, there is presumably a certain degree of belief p'_b, however small, that may rationally be held in my prediction. *The evidence as well as the prediction must be considered.*

This simple example shows how one may make a perfectly definite scientific statement—one that is meaningful—without conveying much if any knowledge. In fact, I should say that the statement that it is going to rain day after tomorrow, free of any supporting evidence and the source of the statement, conveys no knowledge at all. The results of experimental work may also be summarized in terms of meaningful statements that do not transmit knowledge, in that the one who reads the summary may not know how much belief to place in it. Likewise one may present a set of original data without making any interpretative statements. Hence, in what follows we must consider ways and means for presenting experimental data in three different ways: (*a*) as original data, (*b*) as interpretive predictions, and (*c*) as knowledge.

A statement may convey meaning, yet not convey knowledge

Three ways of presenting experimental data

THE RESULTS OF MEASUREMENT PRESENTED AS ORIGINAL DATA

Presenting data as facts; can it be done? Often the engineer and the scientist look on the problem of presenting experimental data as though its solution were independent of the use to be made of the results presented. They try to present the data as scientific "facts." For example, the "pure" physicist or the "pure" chemist studies the structure and properties of materials; measures the fundamental constants of nature; and seeks to discover the "laws" of nature—all with the idea of learning the facts of nature. Such a scientist may not be concerned with any industrial use of his data. In any case he usually contents himself with the thought that he is "presenting the facts" that any application must start with. This is because it is conceived (e.g.) that the velocity of light c, the gravitational constant G, and all similar constants of nature are objective and independent of the kinds of use that may be made of them. In much this same way, tables of the physical and chemical properties of materials are often treated by both the compiler and the user as though they presented facts. This is quite natural, of course, because one is apt to think of the material contained in such tables as giving the characteristics of an unchanging "real" universe instead of a universe of measurements that are subject to sampling fluctuations.

Original data must be considered as evidence for inferences of various kinds. Rule 1. It is customary practice to conceive of the density of pure iron in any chosen system of units as some *single* value. If such a true value exists and if we could discover it, we could presumably put it down once and for all as a fact. In practice, however, we can not discover this true value; *we can simply make measurements and draw inferences from such measurements about other measurements not yet made if we are to limit ourselves to inferences that can be operationally verified.* As previously stated, knowledge provided by such measurements begins in these measurements and ends in measurements, but does not end in the measurements in which it beings: such knowledge can only be *probable.* Hence we must think of the original data simply as *evidence* for one or more various probable inferences, each of which involves its own specific prediction P. Whenever the original data constitute evidence for some specific prediction P it is usually if not always possible to summarize the original data in such a way that the summary itself constitutes an appreciable fraction of the evidence in the original data. In accord with Postulate II (p. 42), it is assumed that there is an objective degree of rational belief p'_b in the prediction P based upon the original evidence E. By definition we shall say that a summary contains all the evidence in the original data for a specific prediction P when the objective degree of belief based upon the summary is the same as that based upon the original data. In general, a summary does not contain the whole of the evidence in the original data, and what constitutes a good summary of the data for one prediction may not be so good a summary for another.

Since the useful knowledge based upon an original set of data involves a more or less definite set of predictions and associated evidences, we may set down

> ***Rule 1.* Original data should be presented in a way that will preserve the evidence in the original data for all the predictions assumed to be useful.**

But just what predictions will be useful? A human element of choice enters here in much the same way that a human element enters into the choice of the readings of scientific instruments in the laboratory. When one is trying to determine how to present data in a given case, it is essential that he consider the kinds of prediction that may be attempted in the knowing processes to which the data may be subjected. What has just been said illustrates the generally accepted conclusion that *we can not have facts without some theory.*[8]

No facts without theory

Two different problems of presentation—data may or may not arise from statistical control. When presenting data we should differentiate between

(*a*) those that arose under a state of statistical control, and
(*b*) those that did not.

It is necessary to do this because the role played by statistical theory under statistically controlled conditions is fundamentally different from what it is under noncontrolled conditions. For the purpose of the present chapter, I shall assume that drawings from a bowl are in a physical state of statistical control and that the mathematics of distribution theory applies. Likewise I shall assume that measurements taken under presumably the same conditions, *provided*[9] *they satisfy Criterion I of control to the extent of at least 25 samples of four* (p. 37), arise under a physical state of statistical control. As was pointed out in the previous chapter (pp. 70 and 79) the statistician[10] alone is then in a position to set up tolerance ranges and make predictions

For statistical control it is not sufficient that measurements be taken under "presumably the same conditions." They must also satisfy Criterion I

[8] ". . . if there is to be any knowledge at all, *some* knowledge must be a priori." From C. I. Lewis, *Mind and the World-Order* (Scribners, New York, 1929), p. 196.—*Editor.*

[9] For reference to Criterion I, see p. 30. It may be of interest to note that the requirements imposed here and elsewhere in this monograph on the data that are to be treated as though they arose under a physical state of statistical control are operationally much more stringent than the requirements made quite generally, and in particular by Neyman on page 21 of his Washington lectures (previously cited on p. 10). There he classes as random experiments carried out repeatedly with utmost care to keep the conditions constant. Here we require in addition that not less than some fixed number of such measurements must be taken and that these must satisfy a particular criterion. Neyman's requirements would presumably be met by the measurements on the fundamental physical constants (fig. 16, p. 66; fig. 20, p. 82), whereas such measurements do not meet the requirements imposed in this monograph.

[10] Of course, an engineer or scientist may also act as a statistician if he knows statistical theory.

that will have maximum validity. However, under conditions that are not in a state of statistical control, the statistician and scientist must cooperate. In fact, the contribution of the scientist to the use of data as evidence under such conditions may be greater than the contribution of the statistician.

Four important characteristics of original data. There are at least the following four characteristics of original data to be considered in presentation:

Original Data—
1. Numbers representing the numerical values of the measurements.
2. Text describing the condition under which each measurement was made, including a description of the operation of measurement.
3. Human element or observer H.
4. Order in which the numbers were taken.

Thus if we let $X_1, X_2, \cdots, X_i, \cdots, X_n$ represent the numerical values of a set of n measurements of some quality characteristic X, then to every X_i there is some associated condition C_i, an observer H_i, and order i. This situation may be represented schematically by a diagram like this:

$$\begin{array}{ccc} & X_i & \\ \swarrow & & \searrow \\ H_i & & C_i \end{array}$$

Every thoughtful student of science or engineering is aware that each of these aspects of the original data may influence his interpretation of the results obtained in any experiment. Table 5 (p. 69) showing the results of Heyl's measurements of the gravitational constant G, is an example of a tabulation in which the data are divided into three groups corresponding to three different experimental conditions. If the same operator took all the measurements,

$$H_i = H_j \tag{17}$$

If the experimenter is of the opinion that the conditions under which all the n measurements were taken are essentially the same, he would say that

$$C_i \approx C_j \tag{18}$$

In such a situation one might be tempted to overlook the *order* in which the numerical data were taken. However, we have seen in the previous chapters that even under conditions assumed to be the same, order is very important until statistical control has been established, as will be further emphasized below under the discussion of the presentation of the results of measurements from the viewpoint of knowledge.

Order can be disregarded only in controlled experiments

Table 7 is an example of the presentation of a set of 204 observations as original data both in regard to the 204 numbers and the order in which they were taken. The numbers provide the numerical values of 204 measure-

TABLE 7

5045	4635	4700	4650	4640	3940	4570	4560	4450	4500	5075	4500
4350	5100	4600	4170	4335	3700	4570	3075	4450	4770	4925	4850
4350	5450	4100	4255	5000	3650	4855	2965	4850	5150	5075	4930
3975	4635	4410	4170	4615	4445	4160	4080	4450	4850	4925	4700
4290	4720	4180	4375	4215	4000	4325	4080	3635	4700	5250	4890
4430	4810	4790	4175	4275	4845	4125	4425	3635	5000	4915	4625
4485	4565	4790	4550	4275	5000	4100	4300	3635	5000	5600	4425
4285	4410	4340	4450	5000	4560	4340	4430	3900	5000	5075	4135
3980	4065	4895	2855	4615	4700	4575	4840	4340	4700	4450	4190
3925	4565	5750	2920	4735	4310	3875	4840	4340	4500	4215	4080
3645	5190	4740	4375	4215	4310	4050	4310	3665	4840	4325	3690
3760	4725	5000	4375	4700	5000	4050	4185	3775	5075	4665	5050
3300	4640	4895	4355	4700	4575	4685	4570	5000	5000	4615	4625
3685	4640	4255	4090	4700	4700	4685	4700	4850	4770	4615	5150
3463	4895	4170	5000	4700	4430	4430	4440	4775	4570	4500	5250
5200	4790	3850	4335	4095	4850	4300	4850	4500	4925	4765	5000
5100	4845	4445	5000	4095	4850	4690	4125	4770	4775	4500	5000

ments of the insulation resistances of as many different pieces of a new kind of material produced under presumably the same essential conditions. The order in which the test pieces were made is that obtained by reading from the top down in each column beginning at the left of the table. These are the data shown in the control charts of fig. 7 (p. 32), where they were considered first in the order in which they were taken and then without respect to this order. There we found that it was the order that furnished the clue to the presence of assignable causes of variability that were later found and removed.

Summarizing original data; by symmetric functions; by Tchebycheff's theorem. Rule 2. It is well to keep in mind that numbers and order are the two aspects of original data that are amenable to mathematical analysis. For example, in any physical or engineering paper there are usually many pages of text descriptive of the operations of measurement and the conditions under which the data were taken. Needless to say, such information is often of very great value as evidence for certain predictions, but there is not, in general, any way available for summarizing it at one stroke. Any *mathematical* summary can present but a portion of what must be considered as the original data. If the order is neglected, as it can be without loss of information when and *only* when the data arise from a state of statistical control, we may present the set of n numbers in the form of an ungrouped frequency distribution. It will be assumed in what follows that such a distribution contains all the useful information in the original set of nonordered numbers. Thus if we have a sample X_1, $X_2, \cdots, X_i, \cdots, X_n$ drawn from a bowl universe (pp. 9 ff), the whole of the

Presentation of results by a frequency distribution—only for controlled experiments

information useful as evidence in predictions is assumed to be contained in the set of numbers arranged in a frequency distribution.

Summarizing by a small number of symmetric functions

To save space, however, it is often desirable to try to summarize a frequency distribution of n finite numbers in terms of a set of m numbers, $\theta_1, \theta_2, \cdots, \theta_m$, where $m < n$ and the θ's are symmetric functions of the n original numbers. The ideal aimed at is to secure a set of these numbers such that one can go from the θ's to the X's as well as he can go from the X's to the θ's. This ideal we may represent schematically as follows:

$$X_1, X_2, \cdots, X_i, \cdots, X_n \rightleftarrows \theta_1, \theta_2, \cdots, \theta_m. \tag{19}$$

Use of Tchebycheff's theorem

Now, of course, without making $m = n$, it is not possible to attain this ideal. It is possible, however, purely from a summary consisting of the sample size n, the average $\bar{X}$ of the n numbers, and their root mean square deviation σ, to say with certainty that not more than n/t^2 of the n numbers $X_1, X_2, \cdots, X_n$ *were* outside the limits $\bar{X} \pm t\sigma$, t being any number whatever that is greater than unity.[11] It should be noted that this statement is true for any set of finite numbers and hence is absolutely independent of whether the set of n data arose under conditions that permit the drawing of valid probable inferences about the expectancy of *future* values of X falling within the range $\bar{X} \pm t\sigma$.

Foı example, let us consider the frequency distribution of the 204 numbers given in table 7. Given only the average $\bar{X} = 4498.18$ and standard deviation $\sigma = 465.24$ oı this set of 204 numbers, one can say with certainty, without ever having seen the original figures, that not more than $204/t^2$ of these numbers were outside the limits $\bar{X} \pm t\sigma = 4498.18 \pm 465.24t$. Tchebycheff's theorem applies as a description of the distribution observed in the sample, irrespective of how the numbers are distributed so long as they are all finite. It is a remarkable theorem, but it does not allow one to differentiate between distributions having the same $\bar{X}$, σ, and n. Hence if the use of the data summarized in the form of the θ's involves inferences that depend on the distribution of the numbers in the sample, it is necessary to give more information in the summary of the original data than is contained in the three statistics $\bar{X}$, σ, and n.

It must be kept in mind that given any prediction P, there is, in accord with Postulate II (p. 42), an objective degree of belief p'_b belonging to the relation between this prediction and the original data. If, instead of starting

[11] *Cf.* Tchebycheff's theorem, W. A. Shewhart, *Economic Control of Quality of Manufactured Product* (Van Nostrand, New York, 1931), p. 95. Tchebycheff's original article "Des valeurs moyennes" (in French) appeared in *Liouville's Journal*, 2d series, vol. xii, pp. 177–184, 1867. A translation into English is given in Smith's *Source Book in Mathematics* (McGraw-Hill, New York, 1929), pp. 580–587. *Editor.*

with the original data, we start with a summary of these data, the corresponding objective degree of belief in the same prediction P may turn out to be somewhat different. Hence we may set down the following rule for summarizing a frequency distribution of data in terms of symmetric functions:

> *Rule 2.* **Any summary of a distribution of numbers in terms of symmetric functions should not give an objective degree of belief in any one of the inferences or predictions to be made therefrom that would cause human action significantly different from what this action would be if the original distribution had been taken as a basis for evidence.**

The Results of Measurement Presented as Meaningful Predictions

Every interpretation involves a prediction. Criterion of meaning. The idea of presenting experimental results as original data is familiar to all of us. However, presentation as a prediction may not be so familiar; in fact some scientists and engineers may prefer to think of only two ways of presenting the results of experimental work, namely, as original data and as an interpretation. Closer examination reveals, however, that every meaningful interpretation involves a prediction.[12]

As a starting point, it may be helpful to note that the statistician may examine and analyze a sample of data from a normal bowl universe and set down estimates of the true average and standard deviation of the universe. Here the sample constitutes the original data and the estimates constitute an interpretation. As another example, a physicist may examine all the original data on the measurement of Planck's constant h, and then state his findings in terms of the customary units in the form

$$h = (6.551 \pm .013)10^{-27}.$$

Here the physicist has presented only an interpretation. As a third example, we shall consider a recent statement by Sir William Bragg in his book, *The Crystalline State* (Oxford, 1925): "The difference between the three principal states, gaseous, liquid, and crystalline—it is better to say crystalline rather than solid—is brought about generally by an alteration in temperature. When the temperature is high enough, the atoms and molecules are endowed with so much individual energy of movement that they lead a more or less independent existence as a gas. When the temperature sinks somewhat, the forces begin to get the upper hand, and the molecules

[12] The editor is reminded of C. I. Lewis' statements, ". . . there is no knowledge of external reality without the anticipation of future experience." ". . . what the concept denotes has always some temporal spread. . . ." "There is no knowledge without interpretation." *Mind and the World-Order* (Scribners, 1929), p. 195.

join up to make a liquid, but not so tightly as to bind neighbours together permanently and in a definite way." Here we have a beautiful interpretation of many measurements of different kinds that are not given in the text. In each of these examples we can easily distinguish between original data on the one hand and the interpretation on the other. Where then does prediction come in?

Let us consider first the quotation about the crystalline state. This certainly predicts what will happen to a gas when the temperature sinks somewhat. One gets a vivid picture of the way the molecules will join up to make a liquid. Here is certainly prediction. Likewise the statement about Planck's constant may lead us to expect that future measurements of this constant will give us somewhat similar results to those quoted above. In much the same way, the statistician's estimate of a universe parameter is a prediction of what he would expect to find the true value to be if he could measure it without error. In each of the three examples there is an element of prediction, and it is this element that helps to make scientific results useful.

Let us look a little more carefully at this element of prediction. Just *what* is predicted in each case? For example, how would one proceed to check the prediction that the molecules will join up to make a liquid when the temperature sinks somewhat? So far as this can be tested experimentally in a quantitative way, the prediction must be translated into terms of measurements of certain kinds that may be expected when the measurement of temperature sinks somewhat. We certainly can not "see" the molecules joining hands when the temperature sinks. Just so soon as we make such a translation to predictions in terms of future measurements, we have to allow for the fact that future measurements of any kind, even though made under presumably the same essential conditions, will not likely all be identical—a fact tacitly recognized in the tabulation above of Planck's constant and in the example of the statistician's estimate of the true value of a parameter. Thus we see that different predictions in terms of future repetitive measurements are of fundamental importance.

We shall now consider how data may be presented in the form of three types of prediction of interest from the viewpoint of *repetitive* measurements. Two of these have already been referred to in connection with figs. 14 and 15; the first, the type of prediction involved in a Student range, we shall here refer to as type P_1; the second, the type of prediction involved in the tolerance range, we shall refer to as type P_2. The third type of prediction, type P_3, is that involved in estimation and will be considered first since it is the one with which the statistician is most familiar. It will be helpful to keep in mind that predictions P_1, P_2, and P_3 are three special

We shall here consider three types of prediction

types schematically represented by P in fig. 21 (p. 86) as constituting a fundamental component of knowledge.[13] It should be noted that the field as here limited excludes consideration of predictions that arise from a scientific theory except in terms of repetitive quantitative measurements. Obviously all scientific predictions must have definite meanings, and we shall accordingly choose the following

> ***Criterion of Meaning:*** **Every sentence in order to have definite scientific meaning must be practically or at least theoretically verifiable as either true or false upon the basis of experimental measurements either practically or theoretically obtainable by carrying out a definite and previously specified operation in the future. The meaning of such a sentence is the method of its verification.**

Our immediate object is to compare the operationally verifiable meanings of these three kinds of prediction from the viewpoint of presenting data. For example, the presentation of data as a Student range $X \pm \Delta X$ constitutes a *symbol* for a definite prediction or expectancy that is fundamentally different from that associated with a presentation of data either as a tolerance range or as an estimate of some parameter.

Prediction involved in estimation—type P_3. "Best" estimates. Both the scientist and the statistician make estimates. For example, the scientist estimates the true values of physical constants and of the qualities of materials and objects. The statistician appears to go through much the same process in estimating from a sample the parameters of an assumed universe. In fact much of modern statistical theory deals with the problem of determining the "best" estimates of parameters. Granted that the statistician can obtain what he chooses to call the best estimates in cases where samples are drawn from experimental bowl universes, just what do such estimates mean in the sense of the criterion of meaning stated above? In answering this question, we shall discover a type of prediction that I designate as P_3. Recognizing this type, we shall be in a position to consider the significance of such estimates from the viewpoint of the scientist and engineer.

We start with a sample $X_1, X_2, \cdots, X_i, \cdots, X_n$ drawn with replacement and thorough mixing from an experimental bowl universe of known functional form $f(X)$ involving s parameters $\lambda'_1, \lambda'_2, \cdots, \lambda'_i, \cdots, \lambda'_s$. The statistician sets for himself the problem of finding from the sample the corresponding best estimates $\varphi_1, \varphi_2, \cdots, \varphi_i, \cdots, \varphi_s$ of these parameters.

[13] There are many kinds of prediction other than types P_1, P_2, and P_3. For example, quite a little was said in chapter I about the prediction that an assignable cause can be found when an observed statistic falls outside Criterion I. However, a discussion of these three types should serve to illustrate the general principles involved in presenting data as a prediction.

Let us consider first what Neyman has called the *best unbiased* estimate.[14] Any estimate φ_i of a parameter λ'_i is called unbiased by Neyman if the expected value of φ_i is equal to λ'_i. Then from among the unbiased estimates, he chooses that one as best which has minimum variance. Let us look at this best unbiased estimate from the viewpoint of our criterion of meaning. For this purpose, let us assume that the distribution in the bowl is as nearly normal as can be made with a reasonable number of chips—let us say 1000. Now it can be shown that the average $\bar{X}$ of a sample of n is what Neyman calls the "best unbiased" estimate of the average X' in the bowl. In what way can one verify the statement that the average $\bar{X}$ of the sample is the best unbiased estimate?

Apparently two operations are involved. If we can in some way find the average X' in the bowl,[15] we can then see whether the mean value of the arithmetic means of N samples of n drawn from this bowl approaches this average in the sense of a statistical limit (ch. I, page 20) as the number N of samples is indefinitely increased; in this way we should determine whether the average of a sample is an unbiased estimate. To investigate the meaning of "best" operationally would be more difficult; it would be necessary to plot for each unbiased method of estimation a distribution of the estimates of X' obtained by that method from the N samples of n; that method whose distributlon has the smallest standard deviation (or roughly, spread) is the "best."

Hence we see that what Neyman calls the best unbiased estimate of a parameter has a pretty definite theoretically verifiable operational meaning. Neyman emphasizes the fact, however, that other criteria of best estimates have been proposed and still others may be developed. The predictive element in each of these would have a meaning different from that described in the previous paragraph, so that, as he says clearly enough, there is no such thing as *the* best estimate: there can be only an estimate that is chosen (or shall we say accepted?) as best by someone. In this sense, the statistician ends his hunt for a best estimate pretty much as the editors of the *International Critical Tables* end their hunt for the true values of physical constants—both accept some estimate even though they do not accept, in general, the same kind of estimate. In one case the statistician does the accepting and in the other case the scientist does the accepting, but it is acceptance in either case. The statistician, however, can give a very definite operational meaning to his choice of estimate.

[14] J. Neyman, *Lectures and Conferences on Mathematical Statistics* (The Graduate School, The Department of Agriculture, Washington, 1938), pp. 127–142.

[15] Of course, if we make up the bowl ourselves, we know the average X' because we build it up to certain specifications, e.g. normal, with a certain mean X' (e.g. 0), a certain standard deviation σ' (e.g. 1), and some convenient class interval (e.g. $0.2\sigma'$).

It should be noted that the statistician does not attempt to make any verifiable prediction about one single estimate; instead, he states his prediction in terms of what is going to happen in a whole sequence of estimates made under conditions specified in the operational meaning of the estimate that he chooses.

Let us see in what sense Neyman's "best unbiased" estimate could be tested in the practical problem of estimating the true value of some physical constant such as the velocity of light. In this case, we presumably can not find the true value as we could find the true average in the experimental bowl; hence we can not verify in a practical way the prediction that the supposedly unbiased estimate of the true value actually approaches the true value in the statistical sense; the process of verification can only be *theoretical.* However, we could compare the variance of the assumed "best" estimate of a true value of a physical constant based upon N samples of n measurements with the variance of another kind of unbiased estimate of the same true value based also upon N samples of n measurements, and we could see which estimate is better by Neyman's criterion.

Now let us consider the difference between presenting data in terms of estimates and presenting them as original data or summaries thereof in terms of symmetric functions. For this purpose, let us assume that we start with a sample $X_1, X_2, \cdots, X_i, \cdots, X_n$ from a bowl universe of unknown functional form $f(X)$. It is important to note that any of the generally accepted statistical methods of estimating a parameter from the sample of n involves two assumptions: (a) the functional form $f(X)$, and (b) the particular type of estimate to be chosen as best, for example, the best unbiased estimate, the maximum likelihood estimate, or that obtained by the use of the Bayes-Laplace theorem. It is particularly important to note that different kinds of estimates do not have the same operationally verifiable predictive meaning; hence to say that φ_i is *an* estimate of a parameter λ'_i is not in itself operationally definite; we need further to know *what* estimate. The predictive meaning of an estimate is clear (provided the reader knows the kind of estimate), even though the functional form $f_k(X)$ that happened to be chosen to represent the universe as a basis for computing the estimate is not known.

Different kinds of estimates have different predictive meanings, yet the predictive meaning of a particular estimate is independent of the form assumed for the universe

It may be helpful to indicate the relations between the original data and any such set of estimates by the following scheme:

$$X_1, X_2, \cdots, X_i, \cdots, X_n \rightarrow \theta_1, \theta_2, \cdots, \theta_i, \cdots, \theta_m \overset{f_k(X)}{\longrightarrow} \varphi_{k1}, \varphi_{k2}, \cdots, \varphi_{ks} \quad (20)$$

wherein the symbol $\rightarrow$ represents one method of going from the sample to the symmetric functions θ_i, and the symbol $f_k(X)$ over the arrow suggests one of the many ways of going from the θ's to a set of estimates. As just noted, we do not need to know the functional form $f_k(X)$ to interpret the predictive meaning of the φ's.

However, we should note that the φ's by themselves *do not constitute a summary* of the original data in the sense that the θ's constitute such a summary because it is possible, if we know the θ's, to make other sets of estimates based upon other assumed forms for $f(X)$, but it is not possible to do this if we start with a set of φ's and do not know the corresponding assumed form of the universe. An estimate φ_k is of the nature of a *symbol* of something that may be experienced in the future, whereas a statistic θ_k is simply a *summary* of a characteristic of previous observations. This illustrates the sense in which the requirements for presenting the results of measurements as a meaningful prediction are different from those for presenting the original data and preserving the information contained therein for all the useful predictions that might be based on them (cf. Rule 1 on p. 88).

Estimates are predictions but not summaries. A statistic is a summary

Prediction involved in the use of the Student range—type P_1. Sometimes the statistician presents his results in terms of what I have called Student ranges (ch. II). For example, .950 ± .715 is one such Student range corresponding to a probability $1 - p' = \frac{1}{2}$. Now let us ask, what is the operationally verifiable meaning of such a range or what kind of verifiable prediction is it a symbol of?

In the first place, it is of interest to recall that any single Student range such as .950 ± .715 considered by itself is not suggestive of a probability interpretation involving an operationally verifiable prediction (see p. 61). Likewise, that a given estimate is the best unbiased one is not a verifiable statement except when this estimate is a member of a class of estimates. Student ranges for a probability of $\frac{1}{2}$ are subject to the following interpretative prediction: If N samples of the same or different sizes be drawn from the same universe or from other universes, and if in each case a Student range for the probability $1 - p'$ be computed, then we may expect to find that $(1 - p')N$ of the ranges thus set up will include the corresponding true universe averages—a kind of prediction illustrated in fig. 14 (p. 59) and discussed at that point.

Some scientists have sought to calculate Student ranges for the physical constants and they have often chosen a probability $1 - p' = \frac{1}{2}$. Student ranges are also frequently used in other fields of scientific investigation; hence the associated type P_1 prediction is of broad general interest in the interpretation of the measurements of published results of science. The fact that a prediction involving a Student range $X \pm \Delta X$ is a probability prediction concerning not this particular range, but rather of a whole sequence of varying ranges should be of equal interest (p. 59). It should be noted, however, that for most scientific measurements, including those of physical properties and constants, such predictions are only theoretically verifiable because we can not discover the corresponding true values.

From the viewpoint of the presentation of the results of measurement in terms of Student ranges that will provide a meaningful prediction of type P_1, it should be noted that we do not need more than can be presented in the form of a range $X \pm \Delta X$, it being understood that this is a Student range for a given probability, e.g. $p_z = \frac{1}{2}$, or $p_z = .95$, as the case may be.

Practical need for clarification of predictive meaning. We should note that interpretations of such ranges by some scientists appear to involve predictions other than that described above as type P_1, though just what else is involved may not always be clear. Thus in a recent paper, Eddington [16] puts the question: suppose I have occasion to use Planck's constant h and that I find the following two determinations recorded in the literature in terms of the appropriate units:

$$h_1 = (6.551 \pm .013)10^{-27}$$
$$h_2 = (6.547 \pm .008)10^{-27}$$

If we assume that these two are to be taken at their face value, which one shall I choose? He argues that the second one is the more useful to him because it limits h to a narrower range and hence will lead to sharper conclusions.[17] I am not sure what the expression "assuming that these are to be taken at their face value" covers. However, let us assume that it implies that in both instances the values of h were calculated by the same method from what the statistician would call random samples of measurements free from constant errors. Now in what experimentally verifiable sense does the second determination h_2 limit h to a narrower range than does h_1? I assume that all will agree that this inference implies something more than would be implied in the statement that both h_1 and h_2 are ranges interpretable as a type P_1 prediction, but I am not sure what more is implied and hence I am not sure how one would set about checking the meaning.

I assume that anyone who is asked to choose between h_1 and h_2 as estimates of h' would want to choose the value of h that might reasonably be

[16] A. S. Eddington, "Notes on the method of least squares," *Proc. Phys. Soc.* (London), vol. 45, pp. 271–287, 1933.

[17] This seems to imply a prediction of type P_1 in the sense that the tabulated range is supposed to include the true value, although the ranges may not have been computed upon the basis of Student's theory.

expected to be the closer to the true value h'. In fact, the physical scientist usually does not want to set up Student ranges that are expected to corral 50 percent of the true values of the physical constants, but instead he wants to set up ranges that center as closely as possible about the true values. From this viewpoint, which of the two ranges should the scientist choose? Some might be tempted to choose the smaller range as Eddington did. This may be as good a rule as any other to follow under the present assumptions and when we only have the ranges given. However, it is not the rule that one would likely follow if he were given the sample sizes n_1 and n_2 corresponding to h_1 and h_2. For example, so far as we can determine solely from the tabulated ranges h_1 and h_2, the two might have been based upon different sets of data both of which had been taken by the same man under presumably the same essential conditions, the only difference being that the numbers of observations n_1 and n_2 were different. If we were given n_1 and n_2, then the desirable practice to follow would be to choose that value of h that was determined from the larger number of observations.

To make this point clear, let us consider the following two Student ranges obtained from two samples drawn with replacement from a normal bowl universe:

$$.3500 \pm .0200$$
$$.0015 \pm .0218$$

Which one of the averages, .3500 or .0015, would you choose as the one closer to the true average of the distribution in the bowl? Suppose one chooses the value .3500 because it is associated with the smaller range. This might be a reasonable choice simply upon the basis of the tabulated ranges, but it would certainly not be the choice if one knew that the sample size for the first range was 4, and for the other range, 1000. Under such conditions I assume that all will agree that the choice would no longer be .3500. If we turn back to fig. 14 we shall see the location of these two ranges in respect to the true value; the first of these ranges is that shown as the second on this figure and the other is the fourth from the last. I trust that enough has been said to indicate the danger of trying to read into a Student type of range a predictive meaning that is not justified.

Prediction involved in the use of the tolerance range—type P_2. In chapter II (pp. 61 and 62) we considered briefly the predictive meaning of a tolerance range. For example, if we were to say that the 90 percent tolerance range for drawings from a given normal bowl universe is $X \pm \Delta X$, the operational meaning of this is that 90 percent of future drawings from the bowl may be expected to fall within this range.

In practice, of course, we have the problem of establishing tolerance ranges upon the basis of a sample. To make our discussion specific, let us assume that we have at our disposal the following sample of four drawn from what we know to be a normal bowl universe:

1.7   0.2   1.4   0.5

For this sample, the average $\bar{X} = .950$; and the standard deviation $\sigma = .619$. Obviously we can not establish the 90 percent tolerance range upon the basis of this sample. The 90 percent estimated tolerance range for a sample of size four in the sense here considered is approximately $\bar{X} \pm 3\sigma$. For this particular sample of four, this estimated range is $.950 \pm 1.857$, and is subject to verification in two senses. In the first place, it is possible to carry out additional drawings from the bowl to see whether 90 percent of these will fall within the stated tolerance limits. As was pointed out in chapter II, we may seldom expect that tolerance ranges established in this way will be found to be correct. This was illustrated in our discussion of fig. 15 where it was pointed out that only by increasing indefinitely the sample size used as a basis for estimating the tolerance range could we expect to get a tolerance range that would prove to be exact. Hence in practice where we must use estimated tolerance ranges, it is always desirable to record the sample size n—in this case four—that was made the basis of the estimated tolerance range.

Now this *estimated tolerance range* for a sample of four is in addition subject to the following operationally verifiable meaning: if we draw N samples of four from the same normal universe and set up tolerance ranges $\bar{X}_i \pm 3\sigma_i$ $(i = 1, 2, 3, \cdots, N)$ for these N samples and if we then compute the corresponding fractions $1 - p_1, 1 - p_2, \cdots, 1 - p_N$ of the total population in the universe included by these ranges, the average of these fractions will approach[18] approximately .9 as a statistical limit as N is increased indefinitely. This statement implies a prediction about the expected areas of the parent population swept out by the ranges $\bar{X}_i \pm 3\sigma_i$ instead of implying a prediction about the expected number of true values included within the ranges $\bar{X}_i \pm .44\sigma$, where $i = 1, 2, \cdots, N$ (see p. 61).

Common characteristics of the predictions. Now we should note three common characteristics of the scientific meanings of these three types of prediction:

> *A*. The meanings permit practical experimental verification only if, as in the case of experimental universes, we can discover the true average and the true universe area swept out by a range.[19] In most practical cases such as measuring physical constants the method of verification can only be theoretical.

[18] This limit of .9 is approximate in the sense that it was determined empirically by taking the average of $1-p$ for 1000 samples of four. W. A. Shewhart, "Note on the probability associated with the error of a single observation," *Journal of Forestry*, vol. xxvi, pp. 600–607, 1928.

B. The meanings are all in terms of a number of samples and not in terms of a single sample. They do not tell us anything about a characteristic (estimate, Student range, or estimated tolerance range) of a single sample except that it is one of a class.

C. From a practical viewpoint, what is perhaps the most important common characteristic of the meanings of these three types of prediction is that the method of verification theoretically involves an indefinitely large number of samples of size n. Hence when it comes to verifying any one of the three types of statistical statement by experiment it is necessary to take many samples of size n and in this way a large sample—a fact that is of great importance when presenting data from the viewpoint of knowledge, as we shall soon see.

Before leaving the subject of presentation of data in the form of a prediction,[20] let us look again to see how this differs from the presentation as original data. The situation is illustrated schematically in fig. 22.

In passing from the original data on the left of fig. 22 to the predictions on the right, the interpreter takes three steps, involving the introduction of assumptions and interpretive constructs; *he adds something to the original data.* When scientific results are presented as predictions they have operationally verifiable meaning in terms of data that may be taken in the future. They do not, however, in themselves convey knowledge.

The Results of Measurement Presented as Knowledge—Ideal Conditions

We take data to acquire knowledge; how to present the results of measurements as knowledge is therefore of outstanding importance. For example, it was stressed earlier in this chapter that the engineer needs more knowledge about properties of raw and fabricated materials in order to set the most efficient tolerance ranges. Engineers are interested in knowing how they can use statistical theory to help them extract the requisite knowledge from available data and to present it in a form that will be useful to others.

To every prediction there corresponds a certain degree of rational belief. It is necessary now for us to note more carefully than heretofore how knowledge differs from original data and from predictions. Knowledge,

[19] This, of course, involves the assumption that drawings from an experimental normal bowl universe can be said to be in a state of statistical control that is normal. Even here, of course, verification is practical only in the sense that no matter how many samples we have taken in the process of verification, we can always take one more.

[20] Needless to say, the presentation of data as evidence upon which to base a prediction is an entirely different problem. This problem is discussed in the next section, pp. 110 ff.

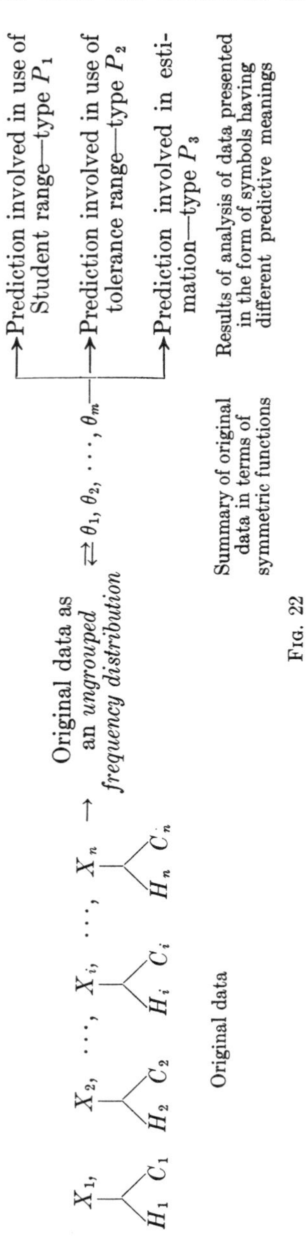
X_1, X_2, ···, X_i, ···, X_n
H_1 C_1 H_2 C_2 H_i C_i H_n C_n
Original data
Original data as an *ungrouped frequency distribution*
$\rightleftarrows \theta_1, \theta_2, \cdots, \theta_m$
Summary of original data in terms of symmetric functions
Prediction involved in use of Student range—type P_1
Prediction involved in use of tolerance range—type P_2
Prediction involved in estimation—type P_3
Results of analysis of data presented in the form of symbols having different predictive meanings

Fig. 22

as has been stated, begins in data and ends in other data. It starts with original data and makes predictions about data not yet taken, involving, at the same time, something more—it involves a certain degree of rational belief in a prediction based upon evidence derived from the original data: this relationship between prediction and evidence is of great importance from the viewpoint of the presentation of the results of measurement as knowledge.

It is perhaps not necessary to point out that just as soon as we begin to consider knowledge it is customary for us to introduce some kind of requirement of truth or validity for the predictions based upon the analysis of the original data. However, the fact is often lost sight of that there is an important distinction between valid prediction in the sense of a prediction being *true*, and valid knowledge in the sense of a prediction being *justifiable* upon the basis of the available evidence and the accepted rules of inference. Thus a prediction may in a given case prove to be false, yet upon the basis of the evidence available at the time the prediction was made, this prediction may be that which the majority of the recognized authorities in the particular field of investigation would have made. From this viewpoint, what might have been acceptable as valid knowledge yesterday may not be acceptable as valid knowledge tomorrow *even though no new data are introduced.*

There is a distinction between a prediction being true, and its being justifiable

For example, the rules of inference accepted by scientists change with time, and as a result what would be accepted today as a valid inference upon the basis of given evidence E might not be accepted tomorrow. Thus an analyst in making predictions of types P_1, P_2, and P_3 (pp. 92–101) makes use of certain distribution theory, and when better distribution theory is developed, the analyst must use it if he is to record the results of measurements, as knowledge, in a way that will be accepted by authorities as valid knowledge. Likewise the scientific analyst must present his evidence along with his predictions if he is to present his results as knowledge. In terms of the schematic diagram of fig. 22, the evidence for the predictions is everything to the left of the three predictions including the deductive as well as inductive chain of reasoning symbolized in the arrows. Such evidence, including the analyst's chain of reasoning, is necessary if the engineer or scientist is to be able to judge the validity of the knowledge solely upon the basis of what is presented.

Knowledge is affected by rules of inference

In all of this, we must remember that it is possible to make statements having a definite operationally verifiable meaning without presenting any evidence (cf. p. 86). Meaning involved in predictions constituting components of knowledge is not only independent of that knowledge but antedates and outruns that knowledge because we must first have a meaning to a prediction before we can decide either its validity or its reasonableness upon the basis of the available evidence; but the meaning of the prediction remains the same even after the validity of the knowledge has been judged.

Nonstatic character of knowledge. Just as soon as we adopt the picture of knowledge here sketched, we are forced to consider knowledge as something that changes as new evidence is provided by more data, or as soon as new predictions are made from the same data by new theories. Knowing in this sense is somewhat of a continuing process, or method, and differs fundamentally in this respect from what it would be if it were possible to attain certainty in the making of predictions. For example, if we had some way of finding out once and for all what the 99.7 percent tolerance range for the density of pure iron is, or what the true value of the velocity of light is (assuming that these things have constant objective values), we could put the figures down once and for all, and they would not change with the acquisition of further measurements.[21] Since, however, we do not know either of these with certainty, and since we can make operationally verifiable predictions only in terms of future observations, it follows that with the acquisition of new data, not only may the magnitudes involved in any prediction change, but also our grounds for belief in it.

Limits to knowing. Predictions based on the bowl universe have maximum validity. The more we know, the more able we are to make valid predictions. Knowledge in this sense is a *process* or a *method of approximating* a practical ideal of a minimum number of false predictions. So far as the three kinds of prediction (pp. 92–101) are concerned, the limiting situation is that conceived of as a state of statistical control represented empirically by drawings from an experimental bowl universe. In fact, it is assumed that if we knew the distribution in the bowl, the validity of the predictions that we could make concerning the fluctuations in the observed characteristics of samples therefrom would represent the limit to which we could hope to go. This is a second characteristic of drawing numbered chips from a bowl universe (with replacement, stirring, etc.), attention having already been directed to the fact that in the state of statistical control so represented, the order and the observer do not constitute useful items of data.

[21] We can, of course, attain this kind of fixity in a deductive science such as mathematics. Deductive statements are either right or wrong and may be verified once and for all by using the conventional formal rules. For example, let us consider the statement that

$$|\pi - 3.14159265358979323846264338328 0| < 10^{-30}$$

We can be sure that such a statement is true once and for all, or is false once and for all.

The object of a scientific investigation and the presentation of its results. It is important to note that statistical distribution theory provides the fundamental basis for predictions in this limiting case. Knowing is a process by which we may hope to approximate closer and closer to this ideal state. I take it that the object of a scientific investigation is so to organize past experience and so to direct the acquisition of new experience that it will be possible to make valid predictions on the outcome of any proposed experiment *that is capable of being carried out,* and to make the prediction in less time than it would take to carry out the proposed experiment. For this reason, the distribution theory of statistics is thus the tool that must ultimately be used for making the kinds of prediction considered here.

Since knowing is of the nature of a developing process directed toward the attainment of an idealized state where maximum validity of prediction results, perhaps the most important requirement on the presentation of data is that *the results of an experiment should be presented in a way to contribute most readily to the development of the knowing process.* This is particularly important in connection with the making of a running report on the quality of product turned out by a repetitive process in mass production where the ultimate goal from the viewpoint of establishing efficient tolerances is the establishment of sets of tolerances having the maximum degree of validity.

The presentation of results from the normal bowl. Let us assume that we know that the distribution in the bowl is normal but that we do not know the two parameters—the average $\bar{X}'$ and the standard deviation σ'. First, let us illustrate a fact to which attention has been called above, namely, that knowing is not static. For this purpose, let us consider the manner of estimating the average $\bar{X}'$ upon the basis of a sample of size n. If we adopt the rule of computing the "best unbiased" estimate (p. 95), we shall find that the estimate for a sample of size $n = n_1$ will not in general be the same as the estimates for $n = n_1 + 1$, $n = n_1 + 2$, etc., as has already been illustrated by the averages computed for the statistical limit in fig. 5 of chapter I (p. 21). Thus knowledge may fluctuate both in the prediction and evidence.

In a similar way, let us consider tolerance range predictions in terms of samples drawn from a normal bowl in which the parameters are unknown. Fig. 15 of chapter II (p. 62) shows how such predictions actually change from one sample of four to another. This same figure, however, does something more—it shows that without a knowledge of the sample size one is not in a position to estimate the size of error that he may make in a prediction of this kind. Hence even in this simple case, we ought to tabulate the sample size if the data are to be used at any time in establishing tolerances. The striking thing, however, is that in the simple case of drawing from a normal bowl, it is sufficient to tabulate merely the average $\bar{X}$, the standard deviation σ, and the sample size n; nothing else is of any use in the predictions here considered.

The presentation of results from a bowl when its distribution is known but is not normal. Theoretically it is feasible to set up ways and means of making valid predictions through the use of statistical distribution theory for each of the three types discussed above (pp. 92–101) for any universe; it is of interest to remark, however, that only a comparatively few functional forms other than the normal law have been investigated. So far as we are here concerned, it is interesting to note that when the known distribution in the bowl is not normal, the tabulation of the average, standard deviation, and sample size will not lead to predictions, particularly of the Student and tolerance range types, that are of the same degree of validity as are those derivable from these same factors when the distribution in the bowl is normal. Furthermore, the answer to the question of how the data may best be summarized under these conditions can be determined with a reasonable amount of labor only by means of theoretical distribution theory such as the theoretical statistician may be expected to supply in the future. For predictions of types P_1 and P_3 it is perhaps reasonable to believe that the average and standard deviation will be two of the symmetric functions that are required. It is almost certain, however, that in order to provide necessary information concerning the magnitude of the errors that may be expected in making tolerance range predictions it will always be necessary to tabulate the sample size n.

Thus for predictions of the three types here considered, it is desirable to tabulate at least the average, standard deviation, and sample size; and for certain non-normal forms of distribution, it is necessary to tabulate other symmetric functions also.

The presentation of results from a bowl when its distribution is unknown. From the viewpoint of presenting the results of measurement, what is different as we pass from the previous case where the functional form of the distribution in the bowl is known, to the present case where it is unknown? The more or less obvious answer is that we must have more information from the sample than in the previous case in order to make the greatest number of valid predictions. It should be noted, however, that so long as the sample of n is drawn from a bowl, it is assumed that the frequency distribution of the numbers in the sample, and the sample size n, contain the whole of the information given by the sample; in other words, it presumably makes no difference who takes the sample and what order is observed.

How shall one make predictions of types P_1, P_2, and P_3 when the functional form of the bowl universe is unknown, particularly if only a small

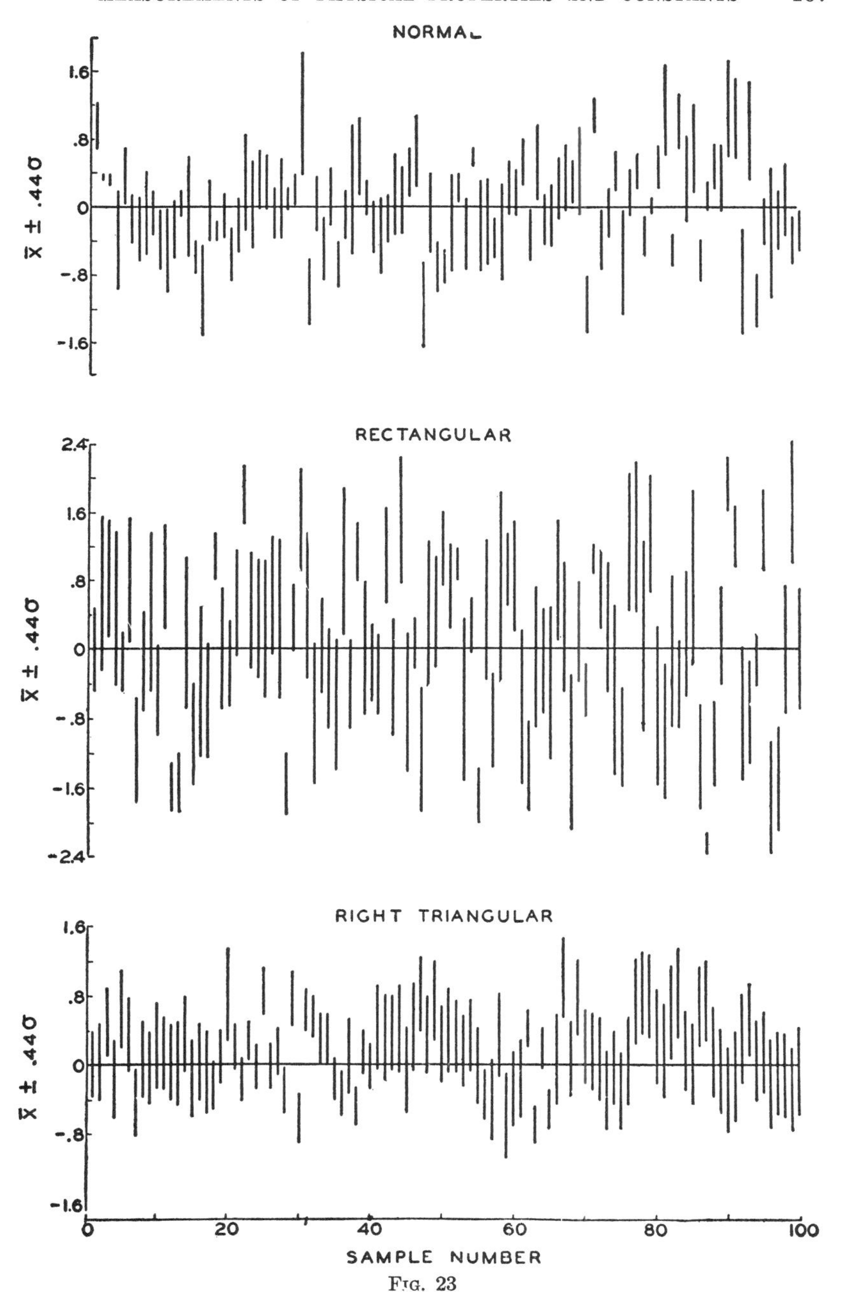
NORMAL
1.6
.8
0
-.8
-1.6
$\bar{X} \pm .44\sigma$
RECTANGULAR
2.4
1.6
.8
0
-.8
-1.6
-2.4
$\bar{X} \pm .44\sigma$
RIGHT TRIANGULAR
1.6
.8
0
-.8
-1.6
$\bar{X} \pm .44\sigma$
0
20
40
60
80
100
SAMPLE NUMBER

FIG. 23

sample is available? What the statistician customarily does is to make predictions as though he were dealing with samples from a normal bowl. Such a procedure may lead to comparatively large errors as a simple example will serve to show. Suppose that one is interested in making predictions in terms of the Student ranges (type P_1) based on samples of four and corresponding to a probability of .50, and that he follows the procedure in setting up such ranges that he would be justified in following if he knew that the samples came from a normal bowl. Fig. 23 shows the results of setting up 100 such ranges corresponding to as many samples of four from each of three different bowls. The functional forms of the experimental universes, although unknown to the observer, were normal, rectangular, and right triangular respectively. Whereas for a normal bowl, 50 of the 100 ranges would be expected to include the true value, the observed number of inclusions for the three sets of data are 51, 56, and 68 respectively. There can be little doubt that the percentage failure of prediction in the rectangular and right triangular cases was largely the result of lack of normality of the unknown parent distribution. This experiment simply illustrates the well-known fact that it is necessary to know the functional form of the distribution in the bowl if we are to attain the limit to which we can go in making valid predictions.

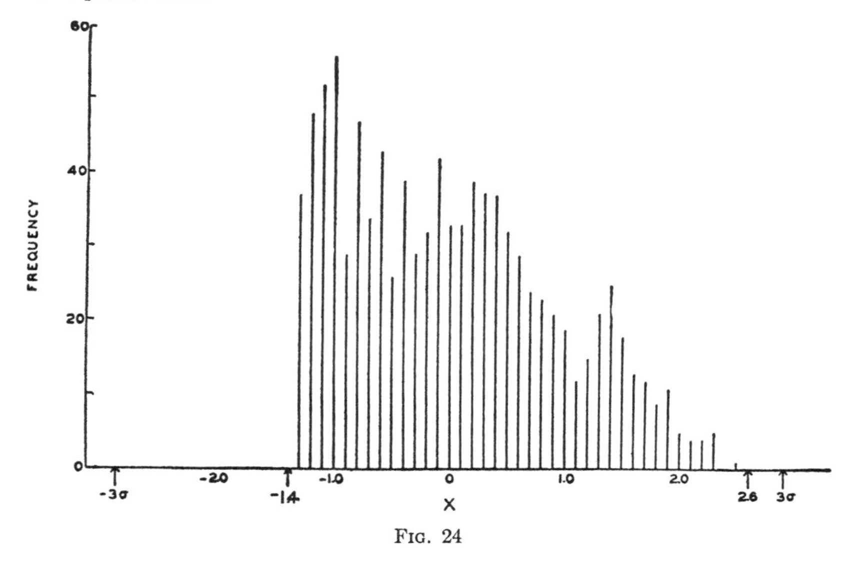

FIG. 24

Although, as we have seen, when the distribution in the bowl is normal, the average, standard deviation, and sample size contain what is perhaps the essential information from the viewpoint of setting tolerance ranges, it is worthy of note that in the present case, where the functional form of the universe is unknown, such a summary is inadequate. Suppose, for example, that we are given $\bar{X} = -.0028$, $\sigma = .9663$, $n = 1000$, and that we are interested in setting tolerance limits for a probability of .997. We have already seen in fig. 15 how accurately tolerance limits can be established upon the basis of such information provided that we know the universe is normal. Suppose now that we set the range for this sample of 1000 in exactly the same way that we would if we knew the universe were normal, by taking the average plus and minus three times the observed standard deviation. This range is laid off on the X axis in fig. 24.

Now let us look at this range in relation to the observed frequency distribution of the sample of one thousand. I think almost everyone will agree that tolerance limits -1.4 and $+2.6$, for example, would satisfy the requirements much more efficiently than the tolerance range $\bar{X} \pm 3\sigma$, which is unnecessarily large. Obviously, in order to go as far as we can in setting valid tolerance ranges, it is essential that we take into account the observed distribution in the most efficient way. In the present state of our knowledge of the theory of estimation and the establishment of valid ranges of variability in terms of a comparatively few symmetric functions, I feel that one is not justified in trying to summarize a sample of the size usually required (1000 or more) solely in terms of symmetric functions as a basis for establishing valid tolerances. We should instead, at least in our present state of knowledge, tabulate the frequency distribution f_0 found in the sample.

As another illustration, let us consider a small sample—a sample of eight drawn from a bowl universe of unknown functional form:

1.7 10.7 0.2 1.4 10.0 10.4 0.5 10.6

How would you summarize these numbers? Would you be satisfied with $\bar{X}$, σ, and n from the viewpoint of predictions of the three types here considered? As a background for answering these questions, let us plot these eight values on a straight line, fig. 25. I believe that it will be generally agreed that the knowledge based on the distribution of numbers, particularly when shown graphically, is sufficiently different from that based on the summary in terms of the average $\bar{X} = 5.69$, standard deviation $\sigma = 4.76$, and sample size $n = 8$ to make it desirable to record the distribution—here, the eight

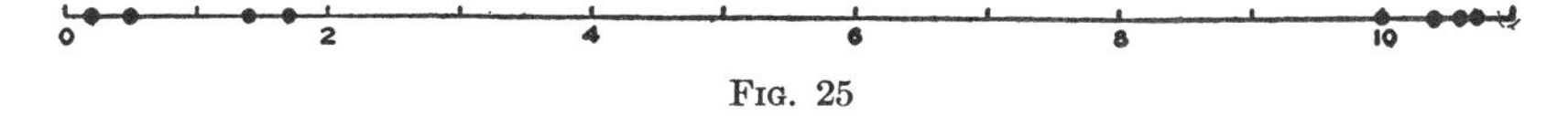

FIG. 25

numbers themselves. For example, suppose one were interested in setting a tolerance range for a probability $1 - p' = .997$ based on this sample. If all one knew was the summary in terms of $\bar{X}$, σ, and n, he would likely set up the range as though the sample had been drawn from a normal bowl universe, and in so doing, he would experience a certain degree of belief in his tolerance range prediction. If now, the same person were shown the original distribution, would his degree of belief be increased or decreased? Mine would be decreased materially. Hence I should want to be given the observed sample distribution f_0 even though the sample size is only 8. Of course, a tolerance range so set would be subject to large error. Anyone familiar with even elementary sampling theory appreciates that a sample of 1000 or more must be available, even when drawing from a bowl, before one can place much reliance in his judgment concerning the functional form of the distribution in the bowl, particularly if one is interested primarily in the tails of the distribution. Furthermore, such a person is familiar with the serious difficulties of trying to judge the form of the distribution when the only information available is a set of symmetric functions such as the θ's in fig. 22, p. 102. So from the viewpoint of summarizing a sample drawn from a bowl in which the form of the distribution is unknown, it does not appear desirable—at least in engineering work and particularly in the setting of tolerances—to give a summary simply in the form of symmetric functions. Hence when tabulating the results that are to be used in setting tolerances when the distribution in the bowl is unknown, it appears desirable to show at least the four quantities

$$f_0, \bar{X}, \sigma, \text{ and } n$$

f_0 being the ungrouped distribution of results in the sample.

The Results of Measurement Presented as Knowledge—Customary Conditions

Complications in real measurements not in a state of statistical control. The problem of presenting the results of measurement of a physical quality characteristic or constant is much more complicated than that considered in the previous section dealing with samples from a bowl universe. This complication arises from the fact that measurements do not in general behave as though they arose under a state of statistical control. In fact, not only do repetitive measurements made by any one method of measuring usually show lack of control, but also measurements of the same quality characteristic or physical constant made by different methods usually indicate the existence of assignable causes of difference. For example, on page 89 we called attention to the fact that for any physical measurement X_i there are three associated elements from the viewpoint of operational meaning, namely the condition C_i under which the observation was taken, the human element H_i introduced by the observer, and the order. Now to assume that an observation such as a drawing from a bowl arises from a state of statistical control implies operationally the assumption that for such an observation we may neglect the factors C_i and H_i as not contributing to knowledge. We shall soon see, however, that these factors play an important role in the problem of presenting the results of physical measurement. How to present C_i and H_i is a difficult problem. It goes without saying, however, that without knowing anything about C_i and H_i there is little ground for believing in any prediction that may be made upon the basis of a series of n repetitive measurements $X_1, X_2, \cdots, X_i, \cdots, X_n$. Here we shall confine our attention to certain aspects of the problem that are significant from the viewpoint of determining the usefulness of statistical theory as a guide in the presentation of the results of measurement.

In the first place, it should be noted in the light of the results presented in chapter II that if the formal rules for making predictions of the three types P_1, P_2, and P_3 are applied to an actual set of physical data, the expectancy of the percentage of valid predictions would be very low compared with the percentage attainable for drawings from a bowl. From the viewpoint of setting the most efficient tolerances, more knowledge is required than is contained in any tolerance set by such a rule, unless we have evidence to indicate that such observations are statistically controlled about a statistical limit which appears to be the same for all of the known methods of measuring. Hence we shall here consider some of the ways statistical theory may be applied to advantage in the process of approaching the idealized condition of statistical control—that is, applied to advantage in the knowing process.

Although there are three component factors of knowledge as here considered, namely *evidence, prediction,* and *degree of belief* p_b, it is noteworthy that we have no quantitative way of measuring p_b. Let us consider the investigations in any new field of measurement. Many, many observations of an exploratory character are often taken before a scientist will even take time to record them. It is almost always a long experimental road between such initial efforts and the announcement of the final results as, for example, in the measurement of the velocity of light by Michelson. For our present purpose, perhaps the most important characteristic of such an approach to scientific knowledge is the fact that *the method of increasing knowledge does not consist in taking more and more repetitive measurements under presumably the same conditions as it does when one is making drawings from a bowl.* In fact, a scientist seldom bothers to take more than five or ten observations under what he considers to be the same essential conditions (drawings from a bowl), although often he experiments with what he thinks may be slightly

Prior to the attainment of statistical control, our knowledge does not increase indefinitely as more and more measurements are taken

different conditions. An illustration is provided by Heyl's measurements of G shown in table 5 (p. 69) wherein the results are given for three different experimental arrangements which we might call conditions C_1, C_2, and C_3, provided it is permissible to conclude that the conditions remain the same for the measurements in each one of the three columns.

Consistency between different methods more important than consistency in repetition. The degree of belief that a scientist holds in a prediction made upon the basis of measurements of some physical constant or property depends a lot more on the consistency between the results obtained under slightly different conditions and by different methods of measurement than it depends upon the number of repetitions made under what he considers to be the same essential conditions. In all such work it has long been recognized that the statistician may contribute to the efforts of the scientist in discovering assignable differences between two or more sets of observations. For example, in table 5 the statistician might apply tests for determining whether the data obtained under the three possibly different conditions could reasonably have occurred as a result of sampling fluctuations; all he needs for this purpose are the average, standard deviation, and sample size for each of the three sets of measurements.

A word on the detection of constant errors by "tests of significance." It is very difficult, however, to weigh the importance of this contribution of the statistician and to determine how much the results of his efforts contribute to a rational belief in the conclusion derived from the analysis of data. From the viewpoint of scientific inquiry, the validity attainable in predictions depends so much upon the skill of the experimentalist in selecting appropriate sense data on the one side and connecting principles or conceptual theories on the other, that unless this process is carried out successfully, almost nothing that the statistician contributes is significant. One must not place too much reliance upon the existence or nonexistence of so-called significant differences reached in any statistical test. However, if the scientist is successful in his choice of data and interpretative principles, the results of the application of statistical tests have the value customarily attributed to them and are successful to this extent. Hence in tabulating data from the viewpoint of providing knowledge, it is often desirable that summaries be made in terms of the average, standard deviation, and the sample size for each group of data taken under conditions assumed to be the same by the scientist, instead of summarizing all the data as though it constituted a single sample from a statistically controlled condition.

Need for the attainment of statistical control. What seems to me a very important contribution of statistical theory to scientific methodology comes about when one tries to go further than the scientist customarily goes in looking to see whether repetitive observations made under presumably the same essential conditions satisfy the criteria of control referred to in chapter I. If one is to attain the kind of knowledge that is requisite for establishing the *most efficient* tolerances—the kind that could be established for drawings from a bowl—it is obvious that one must attain a close approximation to a state of statistical control. Furthermore, as I have said before, it is necessary to have a comparatively large sample, usually more than a thousand, as a basis for establishing the tolerance range if one is to keep within practical limits the error in setting such ranges. What I have termed in chapter I the "operation of control" constitutes an operational procedure for attaining this control and the knowledge requisite for establishing such tolerances. This application of statistics is inherently different from that of making the three kinds of prediction P_1, P_2, and P_3 from a single sample, referred to above. In fact, it is the function of the operation of statistical control to help attain with a minimum amount of human effort a state of control wherein we may with reasonable assurance of attaining valid results make these three kinds of prediction as if they were applied to drawings from an experimental bowl.

An interesting characteristic of this operation of attaining knowledge is that to begin with we can not tell how many observations will be required. So long as we find any evidence of lack of control, we can not estimate the degree of belief that we should hold in any prediction made upon the basis of accumulating data. However, this operational procedure of detecting and eliminating assignable causes provides a method of approaching a state of statistical control of a given repetitive operation in a more or less regular manner. So far as the claims for this operational technique are justified, it follows that the *available data should be so tabulated that criteria of control may be applied*, even when the scientist assumes that his data have been taken under the same essential conditions. An illustration of such a presentation is provided in table 7 (p. 90), which gives the 204 observations of insulation resistance in the order in which the pieces were made. For investigating their state of control, the averages and standard deviations of the successive samples of four would have been a satisfactory summary of the original data.

No universe exists until control is established

It will be noted that in the previous section dealing with ideal conditions (pp. 101–110), the recommendation there given was to present the observed results in an ungrouped frequency distribution f_0, but that no such recommendation is made in the present section; here we are not assuming that control exists, but rather we are attempting to attain it or prove it. The reason is obvious; the use of the observed frequency distribution f_0 is to give evidence concerning the nature of the distribution function in the experimental bowl, whereas in the initial stages of investigation, the condition of control has not yet been attained and *there is no universe (bowl) to be discovered.* For example, there would be little if any advantage, so far as I see, in presenting the 204 observations of table 7 as an ungrouped fre-

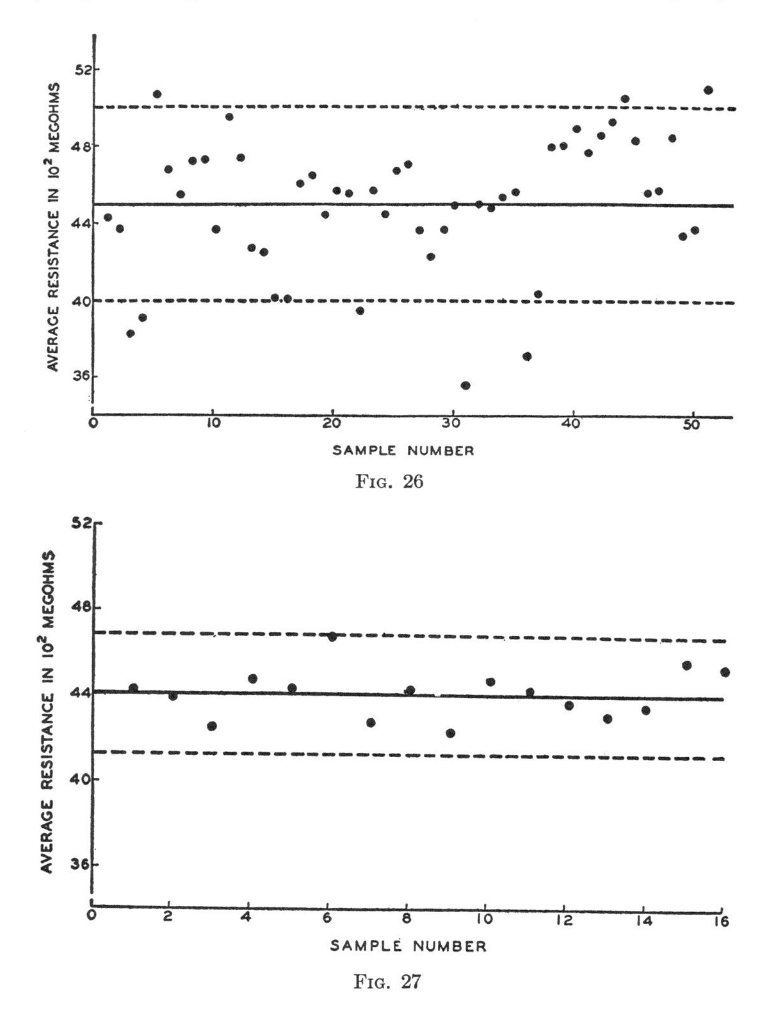

Fig. 26

Fig. 27

quency distribution since these data when tested by means of Criterion I give evidence of lack of control. If, on the other hand, these data had been found to satisfy the criterion of control and if they were to be used as the basis for setting, let us say, a 99.7 percent tolerance range, the observed distribution would be of some help in indicating the validity of such a range.

As we saw earlier (pp. 86 and 103) a prediction devoid of supporting evidence conveys no knowledge. And so it is that in order to convey to another person the knowledge that one obtains from a study of his own experimental work, *it is necessary to present the evidence as well as the prediction.* Since it is customary in experimental work to find that the state of statistical control can be approached only as a limit by discovering and weeding out assignable causes, the presentation of evidence that the assignable causes have been found and removed necessarily adds to one's rational belief that the end results represent a state of statistical control. For example, fig. 26 shows a control chart for 51 averages of four measurements each, derived from a sequence of 204 measurements of resistance on as many pieces of a new kind of product. This figure points with some definiteness to a lack of control, and on this basis certain assignable causes of variability were found and removed from the process, after which the data of fig. 27 were taken. The latter chart gives evidence that it arose from a statistically controlled state, and this belief is strengthened by the recognition that certain causes of variability had been located and removed with the help of fig. 26 before the data of fig. 27 were taken. It is important to keep a running report as a basis for judging quality in mass production because such a report may indicate *progress* toward the attainment of a state of control even though such a state has not yet been attained.

Distinction between summarizing data for evidence of statistical control, and for setting tolerance limits after it has been attained. In the process of testing data for evidence of control, I have shown elsewhere why it is desirable for the scientist or engineer to divide the original data into comparatively small groups which he thinks arose under the same essential conditions. These are then tested for control by some criterion involving in general the use of the average $\bar{X}$, standard deviation σ, and sample size n of each subgroup. Suppose, however, that one wishes to continue the study of the resistance of the new kind of material just considered until he has sufficient evidence for setting valid minimum tolerance limits; beginning with the data shown in fig. 27 and continuing until a sample of something like one thousand or more is reached, the data may be kept in the form of a frequency distribution, for the reason that statistical control may now be assumed to exist. Here we see the difference between (*i*) summarizing data for getting evidence of control, and (*ii*) summarizing data that apparently come from a state of statistical control, for the purpose of providing a basis for establishing tolerance limits that will make possible the most efficient use of material.

Tolerance limits when statistical control has not been attempted. There is, however, another problem that we should consider, namely, that of setting tolerance limits when no attempts at statistical control have been made. In this case, the maximum and minimum observed values play a very significant role in enabling an engineer to set tolerance limits that will include most of the product, although such limits do not permit making the most efficient use of material. This is particularly true if a large number of measurements representing a wide range of conditions is available: the 20,000 measurements of the tensile strength of malleable iron from 17 different sources shown in table 4 of chapter II (p. 65) constitute a good example. For reasons that we need not go into here, the average should also be given, so we may say that under conditions of lack of control, at least the following statistics should be tabulated:

$$\text{Max., Min., } \bar{X}, \text{ and } n.$$

Need for evidence of consistency—constant errors. Let us assume that one wishes to set tolerance limits on the measurement of a physical constant such as the velocity of light. As previously pointed out in chapter II, this problem is the same analytically as that of setting tolerance limits on the true value of quality of pieces of product of a given kind. It is true, of course, that the tolerance limits on a quality must take into account not only the variability of the "true" quality but also that of the method of measurement, hence the problem of setting tolerances on the measurement of a presumably constant value of a given quality always constitutes a part of the job of setting tolerances on a quality characteristic.

Suppose that one is given in the appropriate units the average $\bar{X}$, standard deviation σ, and sample size n for the measurements on the velocity of light previously considered (pp. 67–69):

$$\bar{X} = 299{,}773.85;\ \sigma = 13.37;\ n = 2885$$

Let us also assume, although contrary to fact, that these data satisfy Criterion I of control (p. 30), and that the distribution is approximately normal. Should we be justified in using this set of data *alone* as a basis for setting tolerance limits for the measurement of the velocity of light? Obviously the answer to this question is *No*, if by measurement we are to include measurement not only by the method used in this case but also by other methods admitted by scientists as having a just claim for consideration.

For example, let us compare this set of measurements with another set of 651 more recently reported by Anderson. Fig. 28 shows [22] control charts

[22] Michelson, Pease, and Pearson, "Measurement of the velocity of light in a partial vacuum," *Astrophysical Journal*, vol. 82, pp. 26–61, 1935 (2885.5 observations); W. C. Anderson, "A measurement of the velocity of light," *Rev. Sci. Instruments*, vol. 8, pp. 239–247, 1937 (651 observations).

placed end to end for the two series and constructed as best one can [23] from the data as recorded. The striking thing to note is that the two averages are significantly different. For example, Anderson's data give

$$\bar{X} = 299{,}764.15;\ \sigma = 14.96,\ \text{and}\ n = 651.$$

The ratio of the observed difference in averages to the estimated standard deviation of this difference is

$$\frac{\bar{X}_1 - \bar{X}_2}{.6370} = 15.23.$$

It is indeed very unlikely that a difference so large as this would arise as a result of random sampling. Incidentally, I think that it is this general type

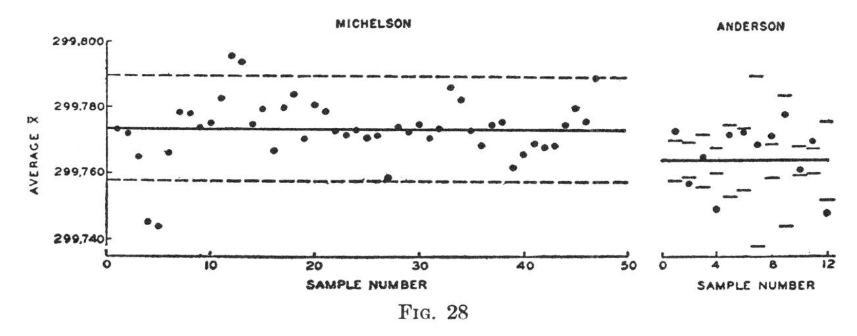

FIG. 28

of experience in which different test methods appear to give assignably different results that leads scientists to stress the importance of looking at the consistency between measurements made by different methods, rather than to stress repetition of the same measurement a great many times (p. 112).

Kinds of information needed for setting limits in uncontrolled conditions. Obviously the kind of evidence that one would want to have before trying to set an efficient tolerance range would be the maximum observation given by the method producing maximum values, and the minimum observation by the method producing minimum values. One would also want to know the number of different methods of measurement that had been tried because "constant errors" have in the past usually been discovered through the use of different methods of measurement. If one takes the time to look back through the literature in physics, let us say for a period of some twenty years or more, he will find quite a variation in the accepted values

[23] Anderson records average deviation for each sample; the sigmas used in the control chart are the mean deviations multiplied by $\sqrt{\pi/2}$. The broken control limits for the Anderson data arise from the fact that his samples are not all the same size.

for many of the constants there tabulated. The same is true for measurements of the atomic weights in chemistry as is illustrated in table 8 which shows the accepted values relative to oxygen = 16 for the dates 1931 and 1936.

From the viewpoint of establishing tolerance limits upon such measurements, it therefore appears that one should record the maximum and minimum values and the number of different methods involved. It would not appear that very much information is provided by a weighted average and an estimate of a so-called probable error so long as the results given by different methods are assignably different. Perhaps in this case more than in any other, the name of the scientist is also an important factor. It would seem, therefore, that statistical theory does not contribute much to the technique of presenting evidence upon which to base a tolerance range under conditions that are not statistically controlled. However, if for some reason it becomes necessary to close up on such a tolerance range by detecting and eliminating all constant errors, statistical tests for significant differences become, as we have seen, a necessary tool in the process.

TABLE 8

INTERNATIONAL ATOMIC WEIGHTS

RELATIVE ATOMIC WEIGHTS

	Oxygen = 16	
Element	*1931*	*1936*
Arsenic	74.93	74.91
Caesium	132.81	132.91
Columbium	93.3	92.91
Iodine	126.932	126.92
Krypton	82.9	83.7
Lanthanum	138.90	138.92
Osmium	190.8	191.5
Potassium	39.10	39.096
Radium	225.97	226.05
Ytterbium	173.5	173.04

Column 2 from table 595, *Smithsonian Physical Tables*, 8th ed. (Washington, 1933)
Column 3 from the *Journal of the American Chemical Society*, vol. 58, p. 547, 1936.

CONCLUDING COMMENTS

We are now in a position to survey in a more critical manner the contribution of statistical methods as they are used for the attainment of knowledge requisite for establishing the most efficient tolerance ranges. The roots of any such knowledge must be grounded in the experience of the scientist. Whatever is selected by the scientist as a basis for quantitative measurement depends upon his intellectual ability in perceiving the important characteristics of a given experience and in making hypotheses and conceptual theories relating these characteristics to others that can be tested by future experiment. The scientist alone is responsible for this initial step. However, when he reaches the stage at which he examines his experiment critically with a view to eliminating assignable causes of variability and attaining a condition of control wherein predictions can be made with the greatest validity, he needs the cooperation of the statistician; it is the statistician who can provide an efficient operational procedure for attaining the state of statistical control. In order that he may apply the statistical technique of control, it is essential that the scientist tabulate the data in such a way that they can be used in the criteria of control. And if knowledge is to be conveyed concerning the attainment or nonattainment of this state, the results of applying control techniques must be presented as evidence.

As soon as a state of statistical control has been attained, the statistician can proceed without the help of the scientist to set up rules that lead to the most efficient prediction. The two may, in other words, part company. Thus we see that the knowing process begins with the scientist and ends with the statistician, but between the beginning and the end the two must cooperate.

Finally, let us ask: What has all this to do with quality control? In the first chapter, we got a picture of the interrelations of the three fundamental steps in control. There, as well as in the second chapter, we saw the need of a running record of quality measurements not only from the viewpoint of giving quality assurance but also from the viewpoint of providing in the end an adequate basis for establishing tolerance limits that will secure the most efficient use of materials, such as is necessary for the establishment of economic standards of quality. In fact an economic standard of quality is not a written finality but is a dynamic process. It is not merely the imprisonment of the past in the form of a specification (Step I, fig. 10, p. 45) but rather the unfolding of the future as revealed in the process of production (Step II) and inspection (Step III), and made available in the running quality report. These facts must be taken into account in the production and exchange of goods if the most economical use of raw materials in the satisfaction of human wants is to be attained. In the preparation of a quality report that will make full use of the additions to available data arising out of a continuing process of mass production, the statistician must play a prominent role.

CHAPTER IV

THE SPECIFICATION OF ACCURACY AND PRECISION

The concept is synonymous with the corresponding set of operations.[1]

P. W. Bridgman, *Harvard University*

Various Aspects of the Problem

Applied science more exacting than pure science regarding accuracy and precision. The development of improved methods of attaining accuracy and precision is an excellent example of the principle that necessity is the mother of invention. When man became a measuring animal he had to adopt standards of length, mass, and the like. Then commerce and industry called for the legalizing of certain standards and the establishment of methods of measuring with ever increasing accuracy and precision in terms of such standards. For example, the introduction of interchangeability about 1787 brought about a need for accurate measurement and the invention of gauges. Then the steady increase in the required accuracy of interchangeable parts produced under manufacturing conditions led to the invention of standard-length gauges with 0.00001 inch tolerances, and pushed the accuracy of test methods out to 0.000001 inch.[2] Both pure and applied science have gradually pushed further and further the requirements for accuracy and precision.

However, applied science, particularly in the mass production of interchangeable parts, is even more exacting than pure science in certain matters of accuracy and precision. For example, a pure scientist makes a series of measurements and upon the basis of these makes what he considers to be the best estimates of accuracy and precision, regardless of how few measurements he may have. He will readily admit that future studies may prove such estimates to be in error. Perhaps all he will claim for them is that they are as good as any reasonable scientist could make upon the basis of the data available at the time the estimates were made. But now let us look at the applied scientist. He knows that if he were to act upon the meagre

[1] *The Logic of Modern Physics* (Macmillan, New York, 1928), p. 5.
The term operation as used by Bridgman is not limited to physical operations but may include in certain contexts what he calls paper and pencil operations and verbalizing. See for example his paper "Operational analysis" in the *Philosophy of Science*, vol. 5, pp. 114–131, 1938.

[2] *Cf.* F. H. Rolt, *Gauges and Fine Measurements* (Macmillan, London, 1929), vol. 1, p. 10.

evidence sometimes available to the pure scientist, he would make the same mistakes as the pure scientist makes in estimates of accuracy and precision. He also knows that through his mistakes someone may lose a lot of money or suffer physical injury, or both.

For example, applied scientists are often called upon to make measurements of many different qualities of raw materials and finished products with *specified degrees* of accuracy and precision. Such specifications may be used in describing some physical quality characteristic of a material to be used in some part of a machine, such as the steering rod of an automobile, or of a material to be used as a food, or of some drug to be used as medicine. In each of these typical cases, a failure to meet the specification may occasion physical injury to someone. In other cases, the specified degrees of precision and accuracy may define conditions economically desirable as, for example, when they define the conditions to be met by the quality characteristics of pieceparts in order that they may fit together in random assembly without an economically prohibitive percentage of rejections. Thus we see why it is that the applied scientist can not stop with *making estimates* of precision and accuracy—he must also *act* on the basis of such estimates. He knows that *this action will reveal his mistakes,* and what is more important, he knows that such mistakes may carry with them serious consequences.

The applied scientist must get enough data

The applied scientist in order to be "successful" can not afford to make too many mistakes even though they be small, and in no case can he afford to make a mistake that is large enough to cause serious trouble. Hence his tendency is to be cautious in accepting any estimate of precision or accuracy as a basis for action. In his language, he wants to be "sure" of his estimates before making them the basis of mass production practices. He does not consider his job simply that of doing the best he can with the available data; it is his job to *get enough data before making his estimates.*

The practical man has yet another worry. He knows that specifications of quality involving requirements of fixed degrees of accuracy and precision may become the basis of contractual agreement, and he knows that any indefiniteness in the meaning of any of the terms used in such a specification, including those of accuracy and precision, may lead to misunderstandings and even to legal action. Hence the applied scientist finds it desirable to go as far as one can reasonably go towards *establishing definite and operationally verifiable meanings for such terms.*

Fivefold objective. More specifically, there are two characteristic kinds of engineering sentences in which the terms accuracy and precision are used. One states a *specification* such as

A. The accuracy of the test method *shall be* ± 1 percent.
B. The precision of the test method *shall be* ± 1 percent.

The other states a *judgment* such as

a. The accuracy of this test method *is* ± 1 percent.
b. The precision of this test method *is* ± percent.

Not only must the engineer clearly distinguish between the meanings of the *concepts* of accuracy and precision but he must also be able to differentiate in an operationally verifiable manner between a specification and a judgment involving either of these concepts; he must also know *what kind* of evidence and *how much* evidence is required as a basis for making a valid statement in the form of a judgment about either accuracy or precision.

It is important to note that, at least in the statement of a judgment (such as *a* or *b*), it is necessary to consider not only the *meaning* but also the *truth content* and the *validity* of the statement. As already noted, a statement in the form of a judgment such as *a* or *b* has the property of being either true or false. This property of a statement or judgment of being either true or false in a specific case must be distinguished from the operationally verifiable meaning of the statement that necessarily antedates and outruns the truth content in a specific case. What is more important, as pointed out in the previous chapter, a judgment always involves a relation between specified evidence and a specified prediction, and the judgment may be valid even though the prediction be false.

A specific example may help to make clear this distinction between the meaning of the prediction involved in a judgment and the validity of that judgment. I have before me a commercial specification for core solder that includes the following chemical requirement: "The percentage of tin shall be determined by any method capable of a precision of ± .2 percent." Presumably this requirement might be stated and might have definite meaning even though no method could be found satisfying the requirement. If, however, I find a method *Z* that I believe meets the requirement, I may upon the basis of specified evidence *E* make the statement, "This method *Z* is capable of a precision of ± .2 percent." This statement may not prove to be true; but whether true or not, the judgment upon the basis of the evidence *E* may still be valid in the sense that it is the judgment that a reasonable man would reach upon the basis of the specific evidence *E*.

The problem in this chapter is fivefold: (*a*) to indicate how far one can hope to go in giving operationally definite meanings to specifications of accuracy and precision; (*b*) to consider available ways and means for determining the truth content of a judgment about either accuracy or precision; (*c*) to consider the operation of determining whether or not a judgment about accuracy or precision is valid; (*d*) to consider ways and means for controlling the error of judgment; and (*e*) to indicate the role played by statistical theory in giving operationally verifiable meanings to statements involving accuracy and precision, and in determining the truth content and validity of such statements.

Broad interest in the problem. Before beginning the technical discussion of this chapter, let us note how many classes of people are interested either directly or indirectly in being able to attain the objectives set forth at the end of the previous paragraph. These objectives are important to the producer and consumer of manufactured goods, both of whom are interested in making specifications that are definite. They are also important for him who would *legislate* a standard of quality in such a way as to minimize the room left for judicial interpretation, and for the court called upon to *adjudiciate* cases arising from such legislation.[3] They are important for every scientist who must record the accuracy and precision of the results of his research, or interpret those of others. That the limit to which we may go in attaining an operationally definite meaning for accuracy and precision constitutes the limit to which we may go in attaining definiteness in any kind of meaning follows at once if we admit, as I think we must, that no greater degree of definiteness is attainable than in the field of quantitative scientific measurement. Hence to provide an operationally definite meaning for accuracy and precision is a fundamental problem for the physical and social scientist, as well as the logician interested in exploring the limits to which one can go in developing an operationally verifiable theory of meaning.[4] Finally it should be of interest to the statistician to discover that statistical concepts and techniques must play a fundamental role in giving definite meaning to the concepts of accuracy and precision as well as in the process of attaining specified degrees thereof in experimental work. Hence the roll call of those interested in the problem considered in this chapter includes

[3] See for example: "Standards and grades of quality for foods and drugs," by Gilbert Sussman and S. R. Gamer, *The University of Chicago Law Review*, vol. 2, No. 4, 1935. The following recent publications of the Chamber of Commerce of the United States are also indicative of the breadth of interest in the problem of legislating standards: *Quality Standards and Grade Leveling*, 1935; *Standardization of Consumers' Goods*, 1934.

[4] Some popular writers have become enthusiastic over the social and scientific advantages that would accrue from increased definiteness in language and have painted in glowing terms the world as it would be if all of us made use of the operational theory of meaning. See, for example, *The Tyranny of Words* by Stuart Chase (Harcourt, Brace and Co., 1937). That there are certain very definite limits to which we may go in attaining the dreamed-of definiteness even in the case of accuracy and precision should be of interest to those who would weigh the importance of such popular expositions.

producers, consumers, scientists, legislators of standards, judges in litigations involving standards, logicians interested in the operational theory of meaning, and lastly, the statistician who is called upon to furnish some of the foundation structure upon which any solution of the problem must rest.

THE MEANING OF ACCURACY AND PRECISION—PRELIMINARY COMMENTS

The terms accuracy and precision have long been and continue to be used by technical people in the discussion of both pure and applied science; they are among those most commonly found in scientific literature. Etymologically the term "accurate" has a Latin origin meaning "to take pains with" and refers to the care bestowed upon a human effort to make such effort what it *ought* to be, and "accuracy" in common dictionary parlance implies freedom from mistakes or exact conformity to *truth.* "Precise," on the other hand, has its origin in a term meaning "cut off, brief, concise"; and "precision" is supposed to imply the property of determinate limitations or of being exactly or sharply defined. Even though there is this definite difference between the etymological meanings of the two terms, they are treated as synonyms in the standard dictionaries and, what is more important, they are often used interchangeably in scientific and engineering literature. In fact, this practice of using the terms loosely and interchangeably has gone to the point where the author [5] of one of the most widely known books on the precision of measurements bemoans the fact that the two terms are so often used carelessly and indiscriminately. Since these terms are frequently used incorrectly and since there is a "rather wide divergence of views in respect to their meanings," [6] they were made the basis recently of a round-table discussion. Such facts are typical of the available evidence indicating that engineers are aware of the existing confusion in the use of these terms at least in some quarters and of the practical need for distinguishing in a definite and verifiable manner between their meanings.

Careful writers in the theory of errors, of course, have always insisted that accuracy involves in some way or other the difference between what is observed and what is true, whereas precision involves the concept of reproducibility of what is observed. Thus Laws, writing on electrical measurements, says: [7] "Every experimenter must form his own estimate of the accuracy, or approach to the absolute truth, obtained by the use of his instruments and processes of measurement. He must remember that a

[5] H. M. Goodwin, *Precision of Measurements and Graphical Methods* (McGraw-Hill, 1913), pp. 7–8.

[6] *Bulletin of the American Society for Testing Materials,* April 1937, p. 23.

[7] Frank A. Laws, *Electrical Measurements* (McGraw-Hill, New York, 1917), p. 593.

high precision, or agreement of the results among themselves, is no indication that the quantity under measurement has been accurately determined." As another example, we may take the following comment from a recent and authoritative treatise on chemical analysis: [8] "The analyst should form the habit of estimating the probable accuracy of his work. It is a common mistake to confuse accuracy and precision. Accuracy is a measure of the degree of correctness. Precision is a measure of reproducibility in the hands of a given operator."

On first reading, these distinctions seem to be clear cut, concise, and to the point. With such distinctions available, why should it have been necessary to hold the round-table conference called by the American Society for Testing Materials in 1937 to consider the meanings of accuracy and precision? With such distinctions recognized in the literature, why at the conclusion of this round table conference was it thought necessary by those present to adopt the following resolution in respect to the word precision? "Resolved that when a standing committee records or specifies a numerical value for precision in a standard, the committee should make clear what is meant in terms of operations or procedures to be followed for purposes of verification." Is it that engineers are not familiar with the literature or is there a more fundamental difficulty? Can it be simply that the cited differences between accuracy and precision are not operationally definite? Let us now look at these differences in a critical manner to see if we can throw any light on such questions.

Why statements about accuracy and precision are often indefinite. Let us note the advice given by Laws to the effect that every experimenter must form his own estimate of the accuracy or approach to the absolute truth. The very phrase "his own estimate" implies that all persons may not be expected to estimate alike. If and so far as different experimenters use different methods for estimating, the advice given by Laws does not have an operationally definite meaning that is the same for all people. Looked at in this way, such advice is indefinite. In the same quotation from Laws, "high precision" is given as synonymous with the phrase "agreement of the results among themselves." But agreement of results among themselves is itself not very definite because there is obviously an indefinitely large number of senses in which results might be said to agree among themselves. For example, in what sense are we to infer that the 204 data of table 7 agree among themselves? We might, for example, think of their agreement in terms of the way they cluster about the observed average or in terms of the magnitude of some one or more of the indefinitely large number of symmetric

[8] Lundell and Hoffman, *Outlines of Methods of Chemical Analysis* (John Wiley and Sons, New York, 1938), p. 220.

functions of these data. Or again we might concern ourselves with the order in which the observations appear.

Much the same kind of indefiniteness exists in the advice quoted from Lundell and Hoffman, wherein accuracy is considered as a measure of the degree of correctness. The meaning of this is definite only if we know what measure is implied and if we know what the degree of correctness is that we are supposed to measure. The phrase "degree of correctness" presumably corresponds more or less with the phrase "approach to the absolute truth" in the advice given by Laws. Likewise, the suggestion that precision is a measure of reproducibility is definite only if we know what measure is implied and if we know what the reproducibility is that we are to measure.

Does this mean that the advice given in the two quotations cited above is not good advice? Quite the contrary. In my estimation at least, it is some of the best advice that I have seen in any practical book discussing accuracy and precision of measurements. Anyone who reads this advice with as much care and thought as the authors apparently used in giving it will get a very definite feeling that accuracy and precision are distinctly different concepts, even though they may not be able to put their fingers on the difference. Furthermore, anyone who reads the books from which these quotations are cited will see that the authors go about the measurement of accuracy in a different way from that in which they go about measuring precision. The point that I wish to make here is simply that such advice is not nearly as definite as we sometimes feel that it is; and furthermore that *it does not provide meanings of accuracy and precision that are subject to experimental verification* as is so often desirable, particularly when such terms appear in specifications that form the basis of contractual agreements.

To emphasize this point, suppose we consider the requirements in a specification that the *accuracy* of the test method shall be $\pm$ 1 percent and that the *precision* of the same test method shall be $\pm$ 1 percent. If there were one and only one experimental method of measuring accuracy, and similarly, one and only one method of measuring precision, and if successive measurements of either accuracy or precision always gave the same identical results, there would be no uncertainty about whether or not in a given case the specification had been met. Just so long, however, as it is admitted (as it seems to be in the quoted advice) that there is more than one way of measuring both accuracy and precision, and just so long as we know that repetitive measurements of either accuracy or precision may not give the same result, such a requirement in a specification loses much of the definiteness that it at first seems to have.

There is yet another and more fundamental sense in which the advice quoted above is indefinite. Most operations of measuring a physical quality characteristic may be repeated again and again an indefinitely large number of times. Such a method may be thought of as being potentially capable of generating an infinite sequence. To what portion of such a sequence do such phrases as "agreement of the results among themselves" or the "reproducibility of the observed values" refer? Unless this question can be answered, the meaning of such phrases is indefinite even though we knew what measure was to be used and what aspect of agreement or reproducibility was to be measured. We shall return later to this point.

In this chapter we are trying to see how far it is possible to go toward making definite statements in the form of either specifications or judgments involving the terms accuracy and precision. Our next step will be to examine briefly the concepts of accuracy and precision as revealed in the theory of errors to see if they provide some of the definiteness lacking in the advice quoted above from practical treatises on measurement.

Accuracy and precision in the theory of errors. Customary assumptions. Let us start with the consideration of what is usually admitted to be the simplest kind of physical measurement, namely, that of the length of the

$$A\text{―――――――――――――}B$$

line AB. To be definite, let us specify that this measurement is to be made with an engineer's scale graduated to 0.01″.

In the theory of errors, we customarily assume that we may repeat such a measurement again and again at will, obtaining an infinite sequence of observations

$$X_1, X_2, \cdots, X_i, \cdots, X_n, X_{n+1}, \cdots, X_{n+j}, \cdots \tag{3}$$

The next step is to assume that the line AB has a true length X' which is constant for all time. Then we introduce the concept of an error e'_i of a single observation X_i defined by the relation

$$e'_i = X_i - X' \tag{21}$$

Thus far everything seems to run along very smoothly.

Now let us ask ourselves, what is the meaning of the accuracy of the method of measuring the length of the line AB by means of an engineer's scale? Of course, one of the things that is done in the theory of errors is to assume that the infinite sequence (3) (p. 12) has a limiting average value $\bar{X}'$; then we sometimes speak of the difference

$$d' = \bar{X}' - X' \tag{22}$$

as a constant error. This constant error provides a kind of measure of the accuracy of the test method in somewhat the same way that eq. (21) provides a measure of the accuracy of the single observation X_i.

Usually, however, we go further and conceive of the accuracy of a given method of measurement as being determined by the frequency of occurrence of the numbers in an infinite sequence such as (3) within some specified range $X' - L_1$, $X' + L_2$. If, for example, we assume that $L = L_1 = L_2$ so that the range becomes symmetrical about the "true" value X', and if we choose L so that the fraction $1 - p'$ of the terms in the infinite sequence (3) that lie within the range $X' \pm L$ is $\frac{1}{2}$, then the distance L is termed the probable error.[9] It should be noted, of course, that p' in such a case is assumed to be a constant value in much the same sense that the true value X' and the expected value $\bar{X}'$ are assumed to be constant values. We may, as is often done, conceive of the probable error thus defined as a measure of the accuracy of the method of measurement characterized by the infinite sequence (3).

Statisticians and experimentalists realize full well that there is nothing sacred about probable error as thus defined, for example, we might choose limits that would include a fraction $1 - p'$ different from $\frac{1}{2}$. Likewise, there is nothing sacred about making $L_1 = L_2$. It appears, however, that most of our common concepts of accuracy in the theory of errors depend in some way or other upon the frequency of occurrence of the numbers in an *infinite* sequence within a range specified in relation to the true value X'.

Passing now to the concept of precision we see that it seems to differ principally from the concept of accuracy in that the clustering of the numbers in the infinite sequence is measured in terms of the fraction $1 - p'$ of these numbers within the range $\bar{X}' - L$, $\bar{X}' + L$, this range being related to the average $\bar{X}'$ of the infinite sequence (3) instead of the true value X' of the thing being measured.

Mathematically all this is extremely simple. For example, we may postulate (i) that repetition of the process of measuring some objective quality characteristic under essentially the same conditions gives rise to an infinite sequence of numbers, approaching, as n is increased, an average value $\bar{X}'$; (ii) that the quality characteristic being measured has a true value X'; and (iii) that associated with any specified range either in respect to X' or $\bar{X}'$ there is a definite fraction $1 - p'$ of the numbers in the infinite sequence lying within this range. In the terms of such postulates, it is a simple matter to differentiate between the concepts of accuracy and precision.

Some difficulties with the usual theory. However, when we try to *apply* the concepts of accuracy and precision based upon such a set of postulates we run into difficulties. To begin with, the first postulate involves the indefinite requirement that the repetitions be made under the "same essential conditions." Does it therefore follow that the theory of errors is applicable

[9] *Cf.* chapter II.

to any sequence observed under what the experimentalist assumes to be the same essential conditions; or would it be better to seek some formal criterion that may be applied to the observed data? Classic error theory attempted to provide the basis for a formal criterion by imposing the limitation that the distribution of the numbers in an infinite sequence should be normal and that the observations should be made at random. However, it was early realized by statisticians that the requirement of normality might be met to a very high degree of approximation by measurements that contain assignable causes of variation; hence the requirement of normality did not provide a satisfactory basis. It is also obvious that in order to apply the concept of randomness it is necessary to have an operation that describes once and for all the meaning of "random." However, classic error theory does not provide such a meaning.

Practically verifiable statements concerning the objective existence of the length of a line

Our effort to attain a definite operational meaning for accuracy and precision would not end here, however, even if we found such a meaning for random, because the meanings for accuracy and precision thus far given are in terms of the unknown and nonexperienceable true value X', expected value $\bar{X}'$, and fraction $1 - p'$ of the numbers in an infinite sequence within certain limits. In the measurement of the length of a line AB, for example, there is no way of observing any one of these three numbers; instead, all that we can experience quantitatively is a finite number of measurements; and the only kind of practically verifiable statement that we can make about the length of the line in the sense that it may be said to have objective existence is that expressible in terms of *a finite number of measurements not yet made.* To make this point specific, let us consider ten observations on the length of one such line, obtained with an engineer's scale reading to 0.01 inch, the next decimal being estimated (table 9).

TABLE 9

4.000	3.996	3.996	3.990	3.994
3.996	3.994	3.994	3.992	3.992

If we are to keep our feet on the ground and make statements that are subject to practical verification, we must express the meaning of the accuracy and precision of the method in terms of the characteristics of a finite portion of the infinite sequence that this operation of measurement is capable of giving. We must go even further if we are to attain operationally definite meanings for statements (specifications or judgments) about accuracy and precision. It is certainly important that we try to determine in what sense the validity of such judgments can be verified: it is just such judgments that form the basis for action and hence in the engineering field are of fundamental practical importance. As a starting point, it will be necessary to consider the nature of operational meaning more carefully and critically than we have yet done.

Operational Meaning

Operation or method of measurement; two aspects. It is important to realize in what follows that there are two aspects of an operation of measurement; one is quantitative and the other qualitative. One consists of *numbers* or pointer readings such as the observed lengths in n measurements of the length of a line, and the other consists of the *physical manipulations* of physical things by *someone* in accord with instructions that we shall assume to be describable in words constituting a text. A simple example of a text outlining an experimental procedure may be useful at this point to help fix the two aspects of a measurement. For this purpose we shall take the following instruction for the measurement of the surface tension T of a liquid: [10]

> In order to make a direct measurement of the surface tension T, attach a very light wire frame a (Fig. 113) to a delicate helical spring s, and by means of an elevating table b, raise a vessel of liquid till the frame is immersed. Next lower the table carefully by means of a rack and pinion r, until a film forms between the prongs of the frame. Then quickly take the reading of the index i upon the mirror scale m. Before repeating, stir the liquid vigorously by means of a glass rod which has been carefully cleaned in a Bunsen flame. Continue this operation until a number of consistent readings can be obtained. The difference between this reading and that taken when the spring and frame hang freely is, of course, a measure of the force of tension F possessed by the two surfaces of the film. In order to reduce this force to dynes, observe the elongation produced by a known weight of the same order of magnitude as F. Then apply Hooke's Law to determine F accurately in grams. Finally measure the distance l between the vertical wires of the frame a with an ordinary metric scale and calculate T from $T = F/2ab$.

The number obtained as T is an example of what is referred to above as a pointer reading. All the rest of this quoted paragraph describes the physical part of the operation.

First let us note that the physical part of even such a simple operation as measuring the surface tension of a liquid is far from being perfectly definite. To begin, we need only call attention to such phrases as "attach a very light wire frame," "lower the table carefully," and "quickly take the reading." Not only are such phrases vague but they must also be understood in terms of other precautions that the experimenter should take, such as making sure that the wire frame and the vessel containing the liquid are

[10] R. A. Millikan, *Mechanics, Molecular Physics, and Heat* (Ginn and Co., New York, 1903), pp. 195–6.

free from grease. "Being free from grease" in turn is not rigorously definite; to some people it means clean enough to eat on; to the experimental physicist it may in some instances mean baked out at a high temperature under vacuum; etc.; and I assume that all would agree that no amount of effort could make such instructions absolutely definite.

Next, let us note that the operation here under consideration is specified not only in physical terms but also in terms of the numerical results obtained by repeating the operation. For example, we have the sentence: "Continue this operation until a number of consistent readings can be obtained." In other words, the text describing the operation does not say to carry out such and such physical operations and call the result a measurement of T. Instead, it says in effect not to call the result a measurement of T until one has attained a certain degree of *consistency* among the observed values of F and hence among those of T. Although this requirement is not always explicitly stated in specifications of the operation of measurement as it was here, I think it is always implied. Likewise, I think it is always assumed that there can be too much consistency or uniformity among the observed values as, for example, if a large number of measurements of the surface tension of a liquid were found to be identical. What is wanted but not explicitly described is a specific kind and degree of consistency. These facts illustrate an important characteristic of every physical measurement considered as an operation, namely, that *neither the physical nor the numerical aspect of an operation by itself can be taken as a complete description of the operation.*

A requirement on the operation is consistency among the observed data

What has just been said is important for the present discussion in that it shows why the definiteness of a specification of an operation depends upon how successfully the requirements upon *both* the physical and the numerical aspects of an operation have been set forth. Likewise the interpretation of experimental results must take into account both aspects of the operation; failure to meet the requirements for either one may be the source of an error in a judgment based upon the observed results. For example, the failure of the experimenter to keep the wire frame and container free from oil in the measurement of surface tension is a source of error. Likewise, the inability of the experimenter to meet the requirement of consistency is a source of error. Furthermore, it is obvious that a criterion of consistency may be met when the requirements on the physical operation have not. Hence it follows that any conclusion that a statistician may derive from the numbers obtained by repeating an operation of measurement must be considered as only part of the evidence in determining the validity of any judgment based upon such an analysis as evidence.

Consistency and reproducibility. Finally, it should be noted that the advice to repeat the operation of measuring surface tension until a number of consistent readings have been obtained is indefinite in that it does not indicate how many readings shall be taken before applying a test for consistency, nor what kind of test of consistency is to be applied to the numbers or pointer readings. Hence we must conclude that the operation of measurement for surface tension quoted above is somewhat indefinite not only in its physical but also in its numerical aspect. One of the objects of this chapter is to see how far one can go toward improving this situation by providing an operationally definite criterion that preliminary observations must meet before they are to be considered consistent in the sense implied in the instruction cited above.

When are the readings consistent?

Before doing this, however, we must give attention not so much to the consistency of the n observed values already obtained by n repetitions of the operation of measurement as we do to the *reproducibility of the operation* as determined by the numbers in the potentially infinite sequence corresponding to an infinite number of repetitions of this operation. No one would care very much how consistent the first n preliminary observations were if nothing could be validly inferred from this as to what future observations would show. Hence it seems to me that the characteristic of the numerical aspects of an operation that is of greatest practical interest is its *reproducibility within tolerance limits throughout the infinite sequence.* The limit to which we may go in this direction is to attain a state of statistical control. The attempt to attain a certain kind of consistency within the first n observed values is merely a means of attaining reproducibility within limits throughout the whole of the sequence.

What about reproducibility?

A requirement concerning a verifiable statement about precision. Just as soon, however, as we begin to consider the reproducibility of the operation in this sense, we must take into account the *whole* of the potentially infinite sequence in trying to define what we mean in an operationally definite way by the term "reproducible." It should be noted that if we are to give definiteness to a test of consistency of the first n observed numbers in an infinite sequence, only these first n numbers are involved, whereas if we are to give definiteness to the concept of reproducibility of the operation of measurement we must take into account the whole infinite sequence or at least that part of it beyond the first n observed values that we arbitrarily choose to consider. Hence it follows that, since any requirement of consistency placed on the n preliminary observations is but a means of insuring reproducibility, the nature of this requirement of consistency can not be given definite meaning until the criteria of reproducibility have been definitely fixed.

It is this characteristic of reproducibility that must be defined in an operationally definite way when we try to give an operationally definite meaning to precision. As we have already noted, the classical concept of precision is stated in terms of the whole of the infinite sequence, but if we adopt this concept, we can never practically determine the truth content of any statement about precision because it is not practically possible to observe the whole of the infinite sequence. *If we are to make a statement about precision that we can verify in practice, that statement must involve a concept of precision that does not take into account the whole of the infinite sequence.* This leads us to a further consideration of verifiability as a criterion of meaning.

Practical and theoretical verifiability. Suppose it turned out that a statement or judgment that the accuracy or precision in a given case is such and such could never be verified, or that it is not possible to determine whether the prediction involved in such a statement in a specific case is true. Particularly within the last decade or so it has been said by many writers that any such statement, not being verifiable, would be meaningless; and from this viewpoint, a statement about precision that involves the concepts of precision that we have attributed to the classical error theory would be meaningless for the reason that we can not practically observe an infinite sequence. The fact is that if we were to adopt practical verifiability as a criterion of meaning, much of what is written about accuracy and precision would be meaningless.

In chapter III, however, we adopted a criterion of meaning (p. 94) that permits either theoretical or practical verifiability. Fig. 29 shows schematically the portions of an infinite sequence that are subject to practical and theoretical verifiability. In this figure, the number j of terms within the

Previously observed	Practically verifiable	Only theoretically verifiable
$X_1, X_2, \cdots, X_i, \cdots, X_n,$	$X_{n+1}, X_{n+2}, \cdots, X_{n+j},$	$X_{n+j+1}, X_{n+j+2}, \cdots$
Past	← — — — — — — Future — — — — — — →	

Present

FIG. 29

region of practical verifiability is assumed to be finite. No matter how large we make j, so long as it is finite, there is an indefinitely large portion of the infinite sequence that remains subject only to theoretical verifiability. Therefore, in order to say anything that is practically verifiable about an unobserved portion of the potentially infinite sequence after having repeated the operation of measurement n times, it is necessary to frame this statement in such a way that it will involve only the numbers of a finite portion of the infinite sequence. To make such statements definite, we must do three things: (1) specify the number j; (2) define the function or functions of the set of j numbers that are to be computed; and (3) specify for each such function ψ_i the interval $\psi_{i1} \leqslant \psi_i \leqslant \psi_{i2}$ within which the function ψ_i must lie if the statement is to be considered true.

The operational meaning of a quality characteristic. The only way one can experience any quality characteristic quantitatively is by means of an operation of measurement. As already pointed out (p. 72), there are usually several known ways of measuring any such quality characteristic and presumably many as yet unknown but knowable ways. For each method of measurement, there is a physical operation that is observably different from the corresponding physical operation for any of the other methods. *The objectivity of a quality characteristic exists only in the consistency between the indefinitely large number of potentially infinite sequences constituting the numerical aspects of the operations.* Any such quality characteristic is therefore operationally verifiable in a practical sense only for statements confined to finite portions of the infinite sequences arising from the specified methods of measuring the quality under consideration. The region of practical verifiability is schematically shown by the numbers enclosed within the rectangle of fig. 30. For convenience, the number of observations

$$
\begin{array}{l|l|l}
X_{11}, X_{12}, \cdots, X_{1i}, \cdots, X_{1n}, & X_{1,\,n+1}, \cdots, X_{1,\,n+j}, & X_{1,\,n+j+1}, \cdots \\
X_{21}, X_{22}, \cdots, X_{2i}, \cdots, X_{2n}, & X_{2,\,n+1}, \cdots, X_{2,\,n+j}, & X_{2,\,n+j+1}, \cdots \\
\vdots & & \\
X_{k1}, X_{k2}, \cdots, X_{ki}, \cdots, X_{kn}, & X_{k,\,n+1}, \cdots, X_{k,\,n+j}, & X_{k,\,n+j+1}, \cdots \\
\vdots & &
\end{array}
$$

FIG. 30

within the region of practical verifiability has been taken as j in each sequence, though the number of observations taken in each sequence need not be the same. To make any practically verifiable statement about a quality characteristic X, we must do four things: (1) specify each of the k physical operations of measurement that are to be considered; (2) specify the number of terms that is to be considered for each infinite sequence (the terms thus specified are represented schematically within the rectangle of fig. 30); (3) define the function or functions to be computed in terms of the set of observations thus specified; and (4) specify for each such function ψ_i the interval $\psi_{i1} \leqslant \psi_i \leqslant \psi_{i2}$ within which the function ψ_i must lie if the statement is to be considered true.

Having now considered the region of practical verifiability for an operation of measurement and also for a quality characteristic, let us next consider the corresponding theoretical verifiability. This is necessary if we are to trace the connection between the theoretical and practical meanings of accuracy and precision and if we are to indicate the usefulness of both.

Physical and logical aspects of theoretical verifiability. Examples. For our present purpose it is desirable to consider two aspects of theoretical verifiability, namely, *physical* verifiability and *logical* verifiability. An infinite sequence can not be realized in practice, but we can always conceive of making *one more measurement* and thus theoretically of getting as long a sequence as we wish. In this sense, an infinite sequence is physically observable; theoretically we can observe as much of it as we like. In much the same way we can not express $\sqrt{2}$ in our "rational" number system, but we can conceive of coming as close to it as we like; by carrying out more calculation we can always get one more figure.

(i) *The true value of* X'. But now let us consider in contrast the concept of the true value X' of a quality characteristic, for example the length of a line AB, or the velocity of light. I am not able even to conceive of a physical operation of observing or experiencing a true length X'. You may argue that there are ways of measuring the length of a line, by any one of which you may obtain a sequence of observations; you may even argue that the limiting average $\bar{X}'$ is equal to X'. But the physical operation is a method of obtaining $\bar{X}'$, not X'. Whether $\bar{X}' = X'$ we shall never know. The true length X' is the given, unknowable, unapproachable, ineffable.[11] It is removed from the pale of observation as securely as $\sqrt{-1}$ is removed from the realm of real numbers; there is not even the question of approximating $\sqrt{-1}$ with the rational and irrational numbers.

A true value X' is not observable by any physical operation

This does not mean that anyone is not free to conceive of there being a true value X', but simply that I am not able to conceive of a *physical operation* whereby I can observe it. The conception of true length in terms of operations with symbols having logical and mathematical significance is possible, but in terms of physical operations such a conception is not possible.

At this point, someone might suggest that we consider the measurement of the sum of the angles of a triangle. It might be suggested that here we *know* the true value of the sum to be 180° independent of any measurement. We must keep in mind, however, that the claim that the sum of the angles of a triangle is 180° rests only upon the acceptance of a certain set of postulates about abstract geometry as being descriptive of our real world. If we had chosen another well-known [12] set of postulates, the sum would

[11] *Cf.* C. I. Lewis, *Mind and the World-Order* (Scribners, New York, 1929), ch. II.

[12] J. W. Young, *Fundamental Concepts of Algebra and Geometry* (Macmillan, New York, 1911), p. 34.

theoretically be greater than 180° and for still another well-known set of postulates, the sum would theoretically be less than 180°. If there were available some physical operation by which we could determine which, if any, of these sets of postulates were true, we could then consider this operation as establishing the true value X'. It has long been agreed, however, that there is no *physical operation* by which we can determine the truth content of a set of postulates.

(ii) *The expected average* $\bar{X}'$. We may now think of the theoretical sense in which the expected average $\bar{X}'$ of the infinite sequence is verifiable. As already noted, it is always possible to conceive of repeating a physical operation of measurement once more, irrespective of how many observations have already been made, and of computing the average of this much of the potentially infinite sequence. This operation, however, in itself does not provide a method of approaching to within a specified range of $\bar{X}'$ unless we *assume* some limiting process. And if we assume a limiting process, it will not necessarily apply to the observed sequence; hence it would seem that the expected value $\bar{X}'$ of an infinite sequence can be considered as verifiable only in the logical sense.

(iii) *The degree of belief* p'_b. Next let us consider the degree of belief p'_b assumed to exist in a prediction upon the basis of evidence E. Here again there does not seem to be any conceivable physical operation of finding p'_b. Hence it too must be considered as being only logically verifiable.

(iv) *Randomness*. Finally we may ask in what verifiable sense a sequence can be "random." Of course we can start with a concept of an infinite sequence that satisfies certain specified postulates as defining a random sequence. To allow for a wide variety of sequences that may be formed from the same set of numbers and yet be called random, the subsequences that may be formed from the original sequence by *some rule that does not depend upon the magnitude of the terms chosen from the original sequence*, as well as the original sequence itself, are usually assumed to satisfy the same set of postulates. However, as pointed out in chapter I, it is not humanly possible for anyone to write down one such original sequence, nor has anyone succeeded in giving a rule whereby a person can determine whether an *observed* sequence satisfies the given set of postulates. Hence the meaning of any such theoretical approach to the definition of random can only be logically verifiable.

As an example of this purely logical or postulational approach, we might limit the meaning of random to those sequences satisfying the following two requirements: (1) the limit of the average of the first n terms of the sequence shall exist and approach $\bar{X}'$ as n becomes infinite; and (2) the limit of the average of the first n terms of any subsequence, formed from the original sequence by a rule such that the choice of the terms to be included in the subsequence does not depend upon the magnitude of the term, shall exist and approach $\bar{X}'$ as n becomes infinite. Since there is an indefinitely large number of rules for selecting a subsequence that will satisfy the requirements laid down, such a definition of random admits an indefinitely large number of random sequences.

If instead of starting with some simple postulational basis, one attempts to set down a set of criteria that sequences drawn "physically at random" from a bowl should satisfy if they are to meet the conditions of random sampling, he discovers potentially an indefinitely large number of such criteria to be considered, whereas only a few have so far been formulated. Such criteria depend upon the frequency distribution of the statistical universe and in this way are not so general as those considered in the previous paragraph. Even if we admit that there may come a time when all of these criteria can be set down, there would still be the difficulty of even conceiving of a rule or operation of writing down one such sequence that would satisfy all of these criteria. Then, even if we could surmount this difficulty, we should have to devise some rule of getting random subsequences from this original sequence. Any attempt to do this will meet serious and, I believe, unsurmountable difficulties.

In some way or other it is desirable to get a formal definition or logical concept of random that is applicable to any *physical process* admitted as being random. For example, let us suppose that drawings with replacement from a bowl gave what was admitted to be a random sequence. Now it is generally (if not always) assumed that a physically random process *may* give any possible order of the numbers defined by the operation. It is difficult to see how one can even conceive of a set of criteria that will admit all of these sequences as being random.

Put somewhat crudely, the point that I have attempted to illustrate is this: even if it were possible to write down a random sequence defined in terms of abstract postulates, and if we could independently carry out physical processes such as drawing from a bowl, such processes to be called physically random, no one has yet so far as I know even *conceived* of a satisfactory rule or operation of relating the two kinds of sequences in a logical manner. The situation is much like that of a differential equation, in which the symbols are purely formal. We may, however, interpret a certain symbol as some physical quantity such as heat, but there is no a priori unique way of relating such a symbol to the experimental results obtainable by measurement.

Now that we have briefly examined the operational meanings of the true value X' and the expected value $\bar{X}'$, which enter into the classic definitions of accuracy and precision, we are in a position to consider the operational meaning of these terms.

The Operational Meaning of Accuracy and Precision

Some fundamental difficulties. It may be helpful to represent schematically what we have seen to be a fundamental difference between the classic concepts of accuracy and precision (fig. 31). Having chosen a statistically controlled operation of measurement, precision is defined in terms of the

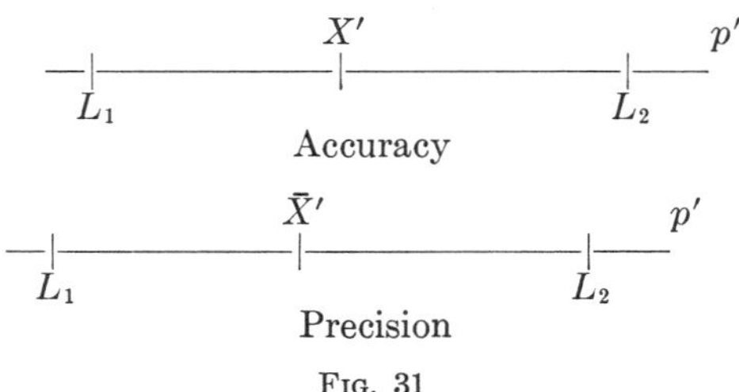

Fig. 31

fraction $1 - p'$ of the numbers in the potentially infinite sequence associated with that operation and lying within a range fixed in respect to the expected or average value $\bar{X}'$ of the sequence. Accuracy for the same operation of measurement essentially differs only in that the range is fixed in respect to the true value X' instead of the expected value $\bar{X}'$. Usually, however, accuracy is thought of in terms of more than one operation of measurement, because the term true value usually implies consistency among all the infinite sequences corresponding to different methods of measurement (cf. sequences (11), p. 72).

The first point I wish to make is that *the ranges used in defining the classical concepts of both accuracy and precision are of the tolerance type in that they are constant ranges conceived of as being tied down to fixed points.*

It should also be noted that the operational meanings of accuracy and precision are more involved than that of either X' or $\bar{X}'$ in that even after these symbols have been given meaning, we must yet consider the operational meaning of p'. Of course, we can logically conceive of the fraction p' associated with any fixed range. However, it is not so easy to conceive of an operation either physical or formal by which one could obtain p' from a *given* observed sequence. In practice, we often think of the fraction p of the first n numbers of an observed sequence and speak of the statistical limit of p as n approaches ∞ as being equal to p'. However, as previously pointed out, this concept of a limit does not provide any formal process of determining how close p approaches p' for any chosen value of n. Hence p' is not formally defined in an operational way other than to say that it may be thought of in much the same way that a true value X' may be thought of, even though we are not able to conceive of an operation of finding either p' or X'.

Enough has been said to show that the symbols p', X', and $\bar{X}'$ entering into the classical definitions of accuracy and precision stand for logical concepts that are neither practically nor physically verifiable. Likewise the concept of degree of rational belief p'_b relating evidence E to a prediction P involving either accuracy or precision is verifiable only in a logical sense. The same situation holds for the logical concept of random. Let us now consider briefly the sense in which accuracy and precision are practically verifiable.

Practically verifiable meaning of accuracy and precision. It follows that the only kind of quantitative and practically verifiable criterion of either accuracy or precision is of the nature of a tolerance range. To make this range operationally definite, we must specify

(1) **the physical operation of measurement for precision, and the one or more such operations for accuracy.**

(2) **the finite number of terms from each of the potentially infinite sequences to be made the basis of the tolerance requirement.**

(3) **the function or functions ψ_i of the terms upon which tolerance limits are to be set, and**

(4) **the tolerance limits ψ_{i1} and ψ_{i2} for each such function ψ_i.**

Here again reference to figs. 29 and 30 (pp. 133 and 134) will be helpful in showing schematically the practically verifiable portion of the infinite sequences that must be used in defining precision and accuracy respectively in terms of measurements not yet taken. These ψ functions may in certain cases be symmetrical functions previously designated by θ, but they are not necessarily so.

These four steps in specifying a tolerance range are of fundamental practical importance in the preparation of operationally definite specifications of quality. It is important to note that we can not speak of *the* practically verifiable meaning of accuracy and precision, but only of *a chosen* verifiable meaning. Furthermore, it appears that such a practically verifiable meaning for either precision or accuracy does not make much use of the concepts of true value X', expected value $\bar{X}'$, fraction p', and random. In the sense that operationally verifiable criteria of accuracy and precision reduce to tolerance range requirements, it is apparent that if one were to stop at this point, he might be misled into thinking that one is free to choose at will any specific verifiable meaning for accuracy and precision. However, when specifying these terms in practice, one is not free to choose arbitrarily *any* conceivable requirement, no matter how much he would like to do so, because he must limit himself to

Tolerance requirements for accuracy and precision must be economic

those that are economically attainable. In other words, tolerance requirements for accuracy and precision must be economic. We have already considered at some length in chapter II the problem of setting such tolerances.

Within this limitation, the meaning of accuracy and precision is perhaps sufficient for use in specifying requirements such as "the accuracy (or precision) shall be 1 percent." However, we must go further in our consideration of meaning if we are to give an operationally verifiable interpretation to a judgment or statement such as "the accuracy (or precision) is 1 percent." We must, in other words, provide an interpretation for the process of determining the validity of such a judgment.

Furthermore, since in specifying accuracy and precision we are tied down to the statement of requirements that can be met, we are not free to ignore the importance of the concepts of true value X', expected value $\bar{X}'$, fraction p', and random, all of which enter into the classic concepts of accuracy and precision. Likewise in the interpretation of a judgment, we must make use of the fundamental concept that any judgment involving a prediction P in terms of either accuracy or precision based upon specific evidence E implies an objective degree of rational belief p'_b.

The meaning of these concepts in use. Up to this point we have considered the logical but not the practically verifiable meaning of these concepts as concepts. Now we must consider their meaning in use. For our present purpose, we should recall that a concept as a concept is an abstract logical *form*. The delineation of such a concept is an act of reason and is independent of any necessary connection with empirical or physically observable data. For example, as already noted (p. 135) we may choose at will a postulational basis for a geometry that will make the sum of the angles in a triangle 180°, more than 180°, or less than 180°. Our choice is independent of any *necessary* connection with the sum of the angles of any real triangle determined by some specified operation of measurement. Now let us recall briefly how a concept is used.

In the first place, the application of a concept as a concept to a particular given experience may be hypothetical. For example, we can say that if the sum of the angles of a real triangle is 180°, and if there exists an operation of measurement that can be repeated an indefinitely large number of times, and if 50 percent of the values in this infinite random sequence lie within the range $180° \pm L$, then L is the 50 percent tolerance range for precision for this particular operation of measurement. Likewise in sampling theory, we constantly use concepts as hypotheses. We say again and again something like the following: If we draw a sample of n at random from a normal population with average $\bar{X}'$ and standard deviation σ', then such and such follows. In fact, the mathematical theory of distribution simply provides us with an indefinitely large number of hypotheses consisting of sets of conditions of the form "If . . . , *then* . . . ." Such is the nature, for example, of tests for statistical significance. Such hypotheses may be formulated at will so long as we conform to the accepted rules of abstract logic. However, they have no *necessary* connection with what is observable.

Abstract concepts serve as guides.
More on the distinction between a statement and a judgment

In the second place, it is a fact (of very great importance in use) that thinking of an abstract concept serves as a guide to the choice of a particular operationally verifiable criterion of experience as a basis for action. Invariably each such practical rule of action, so far as it has been adopted as a result of reasoning, is based upon some abstract concept or set of concepts. There is obviously a very important difference between the hypothetical statement, "If the accuracy (or precision) of a specified test instrument is one percent, accept the instrument," and the judgment constituting a criterion of action, "The accuracy (or precision) of this particular instrument is one percent, hence accept this instrument." Such a choice of criterion of action in each particular instance is an instrumental or pragmatic means of correlating experiences. From this viewpoint, that choice from among all the possible choices that is most useful in correlating experience, is the best. However, best in this sense can be determined only by experiment; it can not be determined by pure reason alone; and it can be determined only by finding out experimentally what kind of action under a specified set of conditions works out more successfully than other kinds of action that have been tried.

Practically verifiable procedures for realizing p', X', $\bar{X}'$, p'_b, and randomness. Distinction between the meanings of concepts and operationally verifiable procedures. In chapters II and III respectively we considered the establishment of economic tolerance ranges and the factors to be considered in determining whether a prediction in terms of a tolerance range and based upon specified evidence E is likely to be true. Now we shall briefly indicate how each of the abstract concepts of p', X', $\bar{X}'$, p'_b, and randomness, has suggested and given rise to the development of operationally verifiable procedures whose usefulness in the field of quality control has been justified by experience. Such operationally verifiable procedures do not constitute the meaning of the concepts for they are not susceptible to such meaningful interpretation. Instead, they are procedures in no way connected with the abstract concepts except that thinking of the one leads to the trial of the other.

Engineers and scientists always want to make valid predictions. But as we have seen, validity of prediction in terms of a tolerance range depends upon the degree of reproducibility of the potentially infinite sequence with respect to this range. We conceive of there being a limit to which we may hope to go in attaining reproducibility, and we conceive of this as being that which is most likely to be correlatable with the abstract concept of random. Under such conditions one must search for a practical criterion of randomness, and the outcome in the theory of quality control has been the development of a practically verifiable operation of control. Hence the meaning of random in *use* considered in this monograph is of the nature of such an operation of control.

The next step in the chain of reasoning is to assume that in those cases where we have eliminated the assignable causes in the practical and definite sense of the theories of quality control, we can find a mathematical probability model upon the basis of which we can make valid predictions. This leads us, for example, to assume that associated with the abstract concept of a certain fraction $1 - p'$ of the numbers in an infinite sequence lying within any specified tolerance range there is an observable number $1 - p$ which we may use in our mathematical model. For example, we may choose to associate $1 - p$ with the simple operation of observing the fraction of the numbers in a finite set of n observations found to lie within some specified range. It is of interest to note that *though this practical range can be fixed, it can not be fixed in respect to either X' or $\bar{X}'$, and hence differs fundamentally in meaning from the conceivable ranges so fixed in the classical concepts of accuracy and precision.*

Naturally in practice we must have some operation of trying to approach closer and closer to what we call the true value X'. In our chain of reasoning we assume that we have an operation of measurement giving a sequence whose expected value $\bar{X}'$ is numerically equal to X', but we have also noted that this is not the same as assuming that operationally X' and $\bar{X}'$ are the same. Then we must find some way that constitutes an attempt to approach $\bar{X}'$. This gives rise to the adoption of Postulate I (p. 22), or in other words, the simple practical rule of operation whereby we *choose* to accept the average $\bar{X}$ of $n + i$ observed values in preference to the average $\bar{X}$ of n observed values. Of course, this rule is not considered applicable in practice until it has been shown that the observed portion of the sequence satisfies the chosen operation of control, and until we have done something that we often describe as eliminating constant errors in the sense now to be considered.

In the measurement of what we assume to be a constant of nature or a property of a physical object, practice is modified by the result of thinking of the abstract concept of an objective value X'. The nearest approach that we can attain to such constancy is in terms of reproducibility in each of the admitted physical operations of measurement, and also in terms of consistency between the results thus obtained. In turn this suggests the use of one or more operationally verifiable statistical tests for significant differences between the results obtained by the different methods of measurement that have been adopted. Such tests constitute practically verifiable criteria for the absence of constant errors.

Now we come to the problem of trying to find an operationally verifiable procedure for determining whether a given judgment in respect to accuracy or precision is valid. There appears to be no way of determining quantitatively even an observed degree of belief p_b. All we can do is determine in an operationally definite manner whether the *action* taken by someone on the basis of specified evidence is the kind that someone else would take upon the basis of the same evidence. This is pretty much the kind of practical procedure that has been adopted in the theory of jurisprudence. The method there followed is in general to take the majority opinion of a specified group of *reasonable* men. Perhaps this procedure is as rational as any other to be followed in determining the validity of a judgment in respect to accuracy and precision based upon specified evidence E. This rests upon the assumption that if every reasonable man could experience *the* objective degree of rational belief p'_b in a prediction P upon the basis of specified evidence E, then all such reasonable men would act in the same way. In turn it is assumed that commonness of action on the part of reasonable men is a practical basis for believing that those acting the same way have experienced the same degree of belief. That is to say, if the objective degree of belief possessed by any person on the basis of evidence E is to be measured by his action, then commonness of action on the part of reasonable men is an arbitrary but convenient basis for defining their measured degrees of belief as being equal. Their objective degrees of rational belief, symbolized by p'_b may or may not be equal; they are unknowable in the same sense that the true value X' is unknowable.

We have now reached the stage where we realize that

$$p', X', \bar{X}', \text{ and } p'_b$$

all turn out to have a common characteristic: each stands for a concept that can not be reduced once and for all to an operationally verifiable meaning. Instead, these concepts serve as fountains of suggested operational meanings from which we must choose in order to talk with definiteness in specific instances.

Need for specifying the minimum quantity of evidence for forming a judgment regarding accuracy and precision. It is important to note, however, that in addition to these operationally verifiable criteria in use corresponding to the different fundamental abstract concepts, we have also called attention again and again throughout this monograph to the need for specifying the minimum quantity of evidence that shall be used as a basis for judging accuracy and precision. Too much emphasis can not be placed upon this requirement if we are to control the error of judgment within practical limits. Without such a requirement one might choose sets of operational criteria that would be satisfied and yet not provide against an occurrence of errors in judgment that are prohibitive from the viewpoint of practice. In fact, the necessity for taking a certain quantity of information as a basis for any important act is the fundamental starting point for the application of all of the five other practical operations in use, namely: (1) those of control, (2) use of probability theory, (3) statistical limit, (4) tests for significance, and (5) majority action of reasonable men corresponding respectively to the abstract concepts of random, p', X', $\bar{X}'$, and p'_b.

The meaning of abstract concepts is not unique. Next we should note that we can not justly refer to *the* meaning in use of the abstract concepts in the classic definitions of accuracy and precision, but instead we can only refer to *a chosen specified set of operationally verifiable meanings*, it being possible to set down an indefinitely large number of different sets of such criteria. The ones that have been discussed and illustrated in the preceding pages have proved successful at least in the field of quality control. In this sense, they are fundamentally *experimental* in character, and are not to be confused with generalized concepts that are not subject to experimental verification. Some one else may find a better set. As time goes on, such criteria in use may be expected to change even though the fundamental abstract concepts were to remain the same. However, the details of the abstract concepts also may be expected to change since there is always an interaction between practical procedures and the associated conceptual background.

Conclusions

We started out in this chapter with a fivefold problem and we may now state our conclusions.

First: *How far can one hope to go in giving operationally definite meanings to specifications of accuracy and precision?*

Since there must always be a physical and a numerical aspect to a quantitative physical operation and since it is not possible to make the requirements on the physical part of the operation rigorously definite, it follows that we can not make a specification of accuracy or precision rigorously definite. We can, of course, place practically verifiable tolerance requirements for accuracy and precision in a specification, but only on the numerical results. However, any such set of requirements represents an arbitrary choice from among an indefinitely large number of possible sets and hence no one of them means the same as *the* requirement: "The accuracy (or precision) of the test method shall be 1 percent" (p. 122).

To the extent that we wish to fix the objective properties of the thing specified in terms of accuracy and precision, we must take account of the idealistic concepts of accuracy and precision in classic error theory. These concepts, however, are not practically verifiable and hence can not be made operationally definite. In fact, there is no necessary relation between the abstract concepts entering into the classic meaning of accuracy and precision and the results of any physical operation. We are not free to choose at will an operationally definite set of criteria if we are at the same time trying to approach as close as possible to meanings that can be used to advantage in practice. In fact, we find that associated with each of the five fundamental concepts entering into the classic definitions of accuracy and precision there are at least as many suggested types of operationally verifiable meanings in use, each of which must be taken into account if we are to attain the practical advantages of specifying accuracy and precision. Since these meanings in use change with experience as well as with the detailed aspects of the abstract concepts, *it is impossible to specify once and for all a satisfactory operationally definite meaning for either accuracy or precision.*

Definite meanings for accuracy and precision can not be specified once and for all

Hence, in many instances, it may be desirable to specify accuracy and precision in terms of formal abstract requirements that suggest operationally verifiable criteria in use. For example, the requirement that the accuracy (or precision) shall be 1 percent, if interpreted in the classic way, is a requirement of this character. It has no necessary connection with experience in a specific instance; but any interpretation of observed data in terms of accuracy or precision will of necessity be shaped in accord with our choice of operationally verifiable criteria that are suggested by the abstract concepts of accuracy and precision.

If, on the other hand, a specification of accuracy or precision is limited to the statement of operationally definite criteria, it is perfectly feasible to meet such criteria without attaining the practical objectives for which the criteria were set. It should also be noted that any statement of accuracy and precision in terms of an operationally definite tolerance range fails to fix any requirements of the *reproducibility* that is so vitally important in understanding the practical usefulness of the concepts of accuracy and precision. Reproducibility here refers to a property of the infinite sequence and therefore involves a portion of it not yet observed. It is often of vital importance to include operationally definite criteria of accuracy and precision in addition to the statements of such a general requirement as "the accuracy (or precision) shall be 1 percent."

Operationally definite criteria should often be included in specifications

Second: *What ways and means are available for determining the truth content of a judgment involving accuracy or precision?*

If the requirements in respect to the numerical aspects of a physical operation are stated in a practically operationally verifiable manner, all we need to do is to carry out the operation thus specified in the requirement and see whether it has been met. Even this simple process, however, is not quite so simple as at first appears, in that the one who is to judge whether a requirement has been met must take into account not only the criteria on the numerical aspects of the operation but also the requirements in respect to the physical operation, which, as we have seen, always must be somewhat indefinite. It follows that a human element must always enter into a judgment of either accuracy or precision even though the numerical requirements are stated in a perfectly definite operationally verifiable manner.

However, there is a more important factor to be taken into account, namely, that a specification is fundamentally the statement of requirements as a means to an end which we idealize in terms of the classic concepts of accuracy and precision. For practical purposes, therefore, there is always left over, beyond any verifiable definite specification, something that we may term the *intent of the requirement*. A simple example may help to make this point clear. We have seen that as a starting point for fixing in a definite manner the significance of accuracy and precision it is necessary to adopt some operation of control. In fact, we considered an example of such a requirement stated in definite terms in connection with the description of an operation of measuring the surface tension of a liquid (p. 130). Since, as has been pointed out, it is not possible to write down all the criteria that should be met in such instances, any specific criterion can be considered a necessary but not a sufficient condition. For example, suppose that we were to adopt as an operation of control the technique involving the use of what I have referred to in this monograph as Criterion I.[13] In fig. 8 (p. 35) we showed a set of one hundred averages of four which satisfies this criterion. Assume, however, that a condition arose where this criterion was implied or stated in the specification and where instead of getting the points shown in fig. 8 we got the succession [14] of points shown in fig. 32. Anyone supposed to judge whether Criterion I had been met would have to answer *Yes* if he considered only the letter of the requirement. However, I think he would have to answer *No* if he were to take into account that the specification was intended to define a condition of randomness which is only partially fixed by any criterion such as Criterion I. In this case such evidence

A specification can not be divorced from the intent of the requirement

[13] Page 309 of my book cited on p. 10.

[14] These points are the first ninety-six averages of samples of four drawn from a normal universe and previously shown in fig. 8. Each group of sixteen averages has been arranged in ascending order of magnitude.

would undoubtedly suggest that at least the intent of the requirement had not been met.

Third: *How may we determine whether a judgment about accuracy or precision is valid?*

An example of a judgment of accuracy or precision might be, as already indicated, a statement of the character: "The accuracy (or precision) of this test method is 1 per cent." Such a statement is a prediction in terms of a tolerance range, and this prediction rests upon certain specified evidence E. The act of judging in this sense is an act of reason relating the evidence E and the prediction P, and is discussed in chapters II and III in connection with the general theory of tolerance ranges.

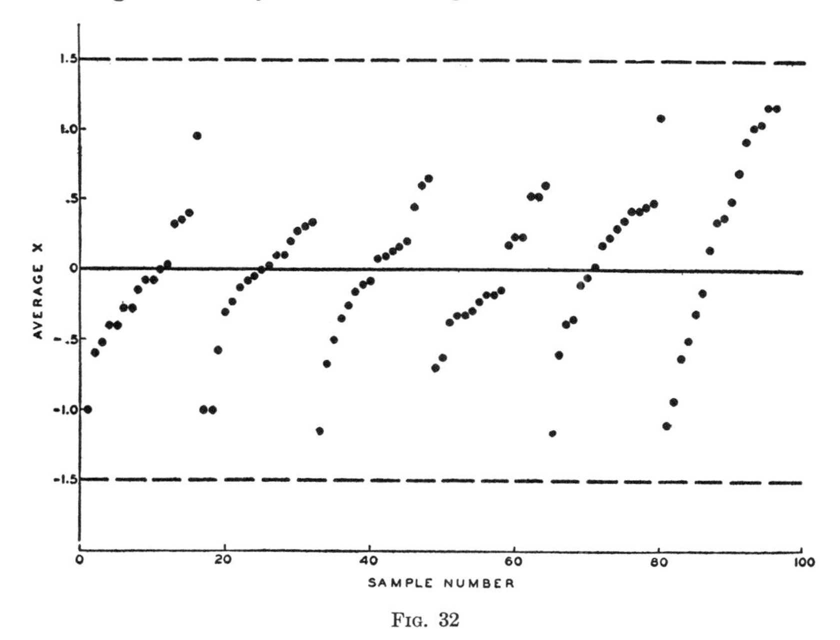

FIG. 32

It should be noted, of course, that such a judgment is a specific act under specific conditions. Its validity is independent of whether the prediction proves to be true and instead depends upon whether the act is the one that a reasonable man should have taken under the particular conditions fixed by the evidence E and the prediction P. It seems that a method for determining operationally whether a judgment is valid is to find out whether it is approved by the majority of reasonable men, where reasonable is taken pretty much in the legal sense of this term.

Fourth: *How may we control the error of judgment?*

The first step in controlling the error of judgment is to take not less than a certain quantity of data, this quantity being fixed by practice in each particular field of inquiry (pp. 37–38). The next step is to make sure that the operation of control is satisfied. We may then proceed to define the desired tolerance type of range for either accuracy or precision. To establish this range we may use the methods discussed in chapter II.

Fifth: *What role does statistical theory play in the specification of accuracy and precision in a definite manner?*

The further we go in trying to fix requirements of accuracy and precision, and in trying to attain in an economic manner a quality of product that will meet such requirements, the more we must rely upon the application of statistical methodology at every step. Starting with the fundamental classic concepts of accuracy and precision, we find that the operations in use associated with these concepts are fundamentally statistical in character. One of the most fundamental requirements underlying accuracy and precision is the reproducibility of an operational procedure, and this leads us at once to the fundamental abstract concept of random and its associated operational verifiable meanings in use. Likewise, in trying to fix the requirement of accuracy corresponding to the concept of an objective true value, we naturally are led to a statement of requirements not only in terms of randomness of a single infinite sequence but also in terms of consistency between sequences as measured in terms of tests for significant differences.

Just as the development of abstract concepts and associated practical techniques go hand in hand in any research, so they have gone hand in hand in the development of ways and means of specifying and attaining accuracy and precision in the control of quality.

EPILOGUE

Hindsight supplements foresight: a view backward often adds materially to a view forward. In his preface, the editor comments briefly on what the reader may expect to find in the four lectures presented in this monograph. A reader who has reached this page can look back and see what he has found. For such a reader, the following paragraphs are offered in the hope that they will help him to round out his picture of statistical method from the viewpoint of quality control.

Central to the theme of the four lectures is the concept of the act of control, which consists of the three components: (*a*) the act of specifying the end to be attained, (*b*) the act of striving to attain the end specified, and (*c*) the act of judging whether the end had been attained. In mass production, as we have seen, these component acts are commonly called specification, production, and inspection. Fundamentally the set of component acts may be put in parallel with three fundamental steps in scientific method as shown below:

Mass Production	*Scientific Method*
Specification	Hypothesis
Production	Experiment
Inspection	Test of hypothesis

The consideration of the three component acts of control as steps in scientific method provides a means of visualizing the act of control as a scientific one, and constitutes a background for the entire discussion in this monograph.

Since the outcome of the repetitive act in mass production, like that of the repetitive one of measurement under the same essential conditions in science, can not be predicted with exactness, we must introduce into scientific method statistical hypothesis, statistical experimentation, and statistical tests of hypotheses. Thus we come naturally to the concept of statistical control. Viewed as an illustration of the role of statistical method in scientific control of the physical world, what is said about the application of statistical theory in the control of quality has an intriguing generality. However, my discussion has been concerned primarily with showing how the theory and practice of statistical control may be made to provide the highest standards of quality of manufactured goods at any given cost. From the practical viewpoint, it is significant that mass production plus statistical techniques when combined in the operation of statistical control provides a continuing, self-correcting process of making the most efficient use of raw materials and fabrication processes. The adjectives continuing and self-correcting are also the essential characteristics of the scientific method.

Chapter I describes the concept of the statistical state of control, the operation of statistical control, and the judgment of control. The assumption that such a state can be attained as a limiting condition in control constitutes the underlying fundamental hypothesis in the theory of statistical control. The five steps in the operation of statistical control provide a practical means of attempting to attain the idealized state (p. 25). Emphasis is placed upon the importance of order in the results of a series of repetitions as a basis for detecting assignable causes of variability. It is shown that the nature of the problem of judging whether a state of statistical control exists is essentially one of testing the hypothesis that assignable causes have been eliminated.

Of fundamental importance for all that is said in this monograph is the fact that the three component acts in the control of quality, namely, specification, production, and inspection, are so interrelated that they can not be taken independently if we are to attain the most efficient control of quality.

Chapter II takes up the very practical problem of establishing tolerance limits that will make possible the most efficient use of raw materials and pieceparts. From a statistical viewpoint, the use of tolerance limits, which are so important in industry, differs in a fundamental way from the use of fiducial limits so extensively discussed in modern statistical theory. Although it is shown that the tolerance range can be reduced toward a minimum with inherent economic advantages as we approach a state of statistical control, evidence is provided to show that such a state is not a natural one, at least in the fields of physical and engineering measurements. This empirically established fact should have some repercussion in many fields where it is the prevalent practice to rest inferences upon the assumption that a state of statistical control (or randomness) exists. It is also of far-reaching significance that even after a state of statistical control has been attained, which is usually a long process in itself, it is still necessary to have available the results of a thousand or more repetitions of the production process if we are to be able to set valid tolerance limits that will provide maximum efficiency in the use of materials. Such facts point to certain inherent advantages of mass production in scientific control.

The problem of providing an efficient running quality report in mass production as discussed in chapter III is obviously of great practical importance. The discussion of this problem leads to a consideration of three aspects of scientific knowledge and in so doing may be suggestive of improved practices to be developed in many other fields of presenting the results of scientific experimentation as "knowledge."

The fourth chapter takes up the simplest type of problem—specifying in an operationally verifiable way a state of statistical control of a single quality characteristic. Such a specification must introduce the concepts of both precision and accuracy. It becomes necessary to make use of and to extend the operational theory of meaning both theoretically and practically to attain the desired end of practical verifiability. Not only is the material here discussed of fundamental importance on account of providing a scientific basis for writing operationally definite specifications of quality, but it may be of considerable interest to statisticians as well as others in attempting to say what they mean and to mean what they say.

Although time did not permit a discussion of the role played by the so-called statistical design of experiments in the control of quality, the importance of the use of such statistical foresight in the layout of the measurements to be made is emphasized in step 2 (p. 25) of the operation of control, and in step 2 (p. 139) of the act of specification.

Throughout this monograph care has been taken to keep in the foreground the distinction between the distribution theory of formal mathematical statistics and the use of such theory in statistical techniques designed to serve some practical end. Distribution theory rests upon the framework of mathematics, whereas the validity of statistical techniques can only be determined empirically. Because of the repetitive character of the mass production process, it is admirably suited as a proving ground wherein to try out the usefulness of proposed techniques. The technique involved in the operation of statistical control has been thoroughly tested and not found wanting, whereas the formal mathematical theory of distribution constitutes a generating plant for new techniques to be tried.

SOME COMMENTS ON SYMBOLS AND NOMENCLATURE

It is a well-established practice of many authors to include a list of symbols used, and at the suggestion of the editor, I undertook to prepare one to be inserted at this point. A start was made by putting down the following description of the symbol X and the symbols $X_1, X_2, \cdots, X_n$.

Symbol	*Description*
X	Some measurable quality characteristic
$X_1, X_2, X_3, \cdots, X_n$	Numbers denoting the results of n observations on some quality X.

A careful reader, however, would immediately point out that X had also been used as a mathematical variable in several different places as, for example, in the equation $dy = f(X)dX$, where dy represents the probability that X will fall within an interval $X \pm \frac{1}{2}dX$. Of course, we may let X represent some quality characteristic, but this does not make a mathematical variable X the same as a quality characteristic X.

In much the same way, $X_1, X_2, X_3, \cdots, X_n$ are not always referred to simply as n numbers denoting the results of observations on some quality X. For example, they are sometimes referred to as a sample, and at other times as measurements. Here we have three different descriptions of the same n symbols. Most of these inconsistencies, if they may be called such, would not likely give much cause for worry. They are, in fact, the kinds of inconsistencies present in most discourse, even in the natural sciences. But now let us pass to a less familiar symbol like X' or $\bar{X}'$. The very inconsistencies that we may be willing to slur over in every day practice are the ones that need to be stressed in learning how to make the best *use* of such terms.

For example, we might describe X' as the mean of a universe or population. Mathematical statisticians have a perfectly definite way of using such a mean in formal mathematics. But what is the meaning of X' in the physical world? Do we have any "statistical universes or populations" in the true sense? The answer to this must involve some consideration of the concept of random operation, and I trust that enough has been said to indicate the difficulties that we get into when trying to describe randomness in an operationally definite way. On the other hand, the usefulness of statistical theory depends on our giving that operation a definite meaning as has been done, for example, in the case of the operation of statistical control.

The same symbol X' is also used in this monograph for the true value of a physical constant, and as such plays an important role in the discussion of errors. In the mathematical theory of errors, the term true value here represented by X' is used consistently by most authors. However, when we try to appraise the usefulness of the mathematical theory involving the use of X' we must think of the objective meaning of X' in the world about us. Such meaning is of paramount importance in the specification of quality, involving as it does the concepts of both accuracy and precision. It is for this reason that operational meanings both theoretically and practically verifiable have been introduced, and a distinction drawn between meaning solely as a prediction and meaning in knowledge. The same type of discussion could be given about the description of every symbol that I have used including not only letter symbols such as X, X', $\bar{X}'$, p, p', and the like, but also word symbols such as random.

What is the trouble with our symbolism? Is it not possible to find a satisfactory one that can be described in a simple but definite manner? Such questions demand consideration. To do justice to these questions would take far more space than is here available. However, I shall try to suggest what appears to me as a helpful manner of approaching the meaning of symbols.

But first let us consider a question that may be in the minds of some readers. Why all this fuss over symbols here when there is not so much fuss in such fields as physics and chemistry, for example? Well, in such fields, the usefulness of mathematical theory has been pretty well established over a long period of research and application. Put somewhat crudely, the physicist and chemist have learned by experience how to extract certain numbers from their experimental work and how to put these into the mills of the mathematician which grind out other numbers or functions that the scientist has learned by experience can be used in certain more or less well-established ways. This means that the mathematician and the scientist in these two fields have grown to have a more or less common ground in language.

However, in the case of mathematical statistics, some scientists and engineers still question the usefulness of all the high-brow theory. They appreciate, for example, that it is one thing to assume that a sample has been drawn at random and another thing to base much reliance upon conclusions that rest upon such assumptions. Then, too, the very concepts such as probability, randomness, universe, statistical limit, and the like are indefinite. All in all we may say that at least in the field of engineering we are only now in the process of learning how to get the right kind of data to put into the refined mill of the mathematical statistician. What is more, engineers know that the usefulness of this mill can be proved only by experience that provides operationally definite meanings in use for the terms now appearing in the formal mathematics.

Now let us return to the problem of providing a definite scientific symbolism. To begin, let us recall that there are at least three important aspects tc every symbol. One of these is the relation of the symbol to the objective thing symbolized; another is the relation of the symbol to the individual or group interpreting the symbol; and the third aspect is the relation of a symbol to other symbols. Schematically we have the following diagram:

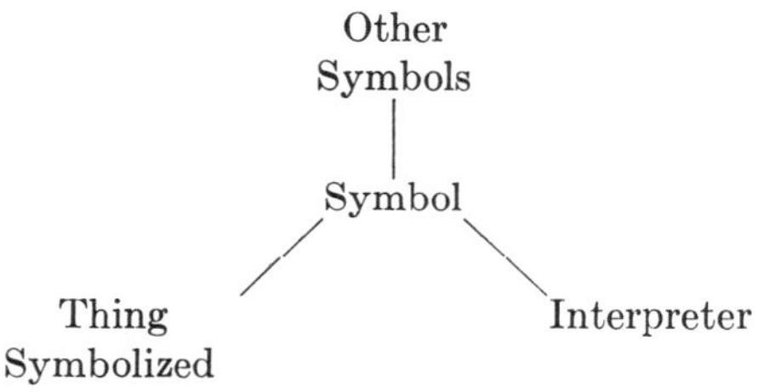

Now let us return to our discussion of the description of the symbol X'. In mathematical statistics this symbol is formally related to other symbols such as, for example, the relation of X' to the symbol for the error e'_i of an observed value X_i, as shown by the equation $e'_i = X' - X_i$. Then there is the response R of a given interpreter to a symbol X' serving as a stimulus S and sometimes indicated by the expression $S \rightarrow R$. Finally there is a way of relating the symbol to an operationally verifiable experience that is presumably independent of any observer, at least to a first degree of approximation. Such, for example, are the theoretically and practically verifiable meanings introduced in this monograph.

As another example, let us take the word random. The mathematical statistician uses this word symbol in a pretty definite way in relation to other symbols in his theorizing. An experimentalist using this word in its relation to his own work is led to expect such and such to be observable when the symbol is used. Here we distinguish the relation of the symbol to the interpreter. However, the practical man has learned all too well that he may be mistaken in his expectations. What he wants to do is to maximize his ability to predict without disappointment. Thus more or less naturally we are led to search for the best way of relating the symbol to an objective kind of observable experience which we may designate by the term random. We want to learn how to distinguish this kind of experience.

In doing this, we must distinguish between meaning considered simply as a prediction and considered in reference to evidence as a basis for a certain degree of rational belief. The significance of the term random is different in each of the three relations considered in this paragraph.

To each one of these three types of relation there belongs a more or less definite set of rules defining that relation. This is particularly true in the case of the relation of a symbol to other symbols as found in mathematical statistics per se. The rules describing the relation of the symbol to the individual are far less definite particularly in the field of applied statistics. Furthermore, those rules in current use are not always the best. From the viewpoint of use, what we particularly want to do is to establish rules for relating symbols to operationally definite and practically verifiable entities that will yield the greatest possible number of valid predictions. For example, I have considered at great length the operational rule of statistical control and the rules for judging when we may assume with a high degree of assurance that a state of statistical control exists. Possibly we are justified in saying that scientifically what we are interested in doing is to establish rules relating the symbols used in mathematical statistics to operationally definite and observable meanings in experience that will lead to the greatest possible number of valid scientific predictions.

In the present state of application of statistical theory in the control of quality, it is essential that we keep in mind each of the three relations characterizing the significance of a symbol.[1]

[1] The reader interested in surveying some of the literature on the significance of signs and symbols in scientific discourse will find a very useful introduction in *Foundations of the Theory of Signs* by C. W. Morris (University of Chicago Press, 1938).

A CATALOG OF SELECTED

DOVER BOOKS

IN SCIENCE AND MATHEMATICS

RELATIVITY, THERMODYNAMICS AND COSMOLOGY, Richard C. Tolman. Landmark study extends thermodynamics to special, general relativity; also applications of relativistic mechanics, thermodynamics to cosmological models. 501pp. 5⅜ × 8½. 65383-8 Pa. $11.95

APPLIED ANALYSIS, Cornelius Lanczos. Classic work on analysis and design of finite processes for approximating solution of analytical problems. Algebraic equations, matrices, harmonic analysis, quadrature methods, much more. 559pp. 5⅜ × 8½. 65656-X Pa. $11.95

SPECIAL RELATIVITY FOR PHYSICISTS, G. Stephenson and C.W. Kilmister. Concise elegant account for nonspecialists. Lorentz transformation, optical and dynamical applications, more. Bibliography. 108pp. 5⅜ × 8½. 65519-9 Pa. $3.95

INTRODUCTION TO ANALYSIS, Maxwell Rosenlicht. Unusually clear, accessible coverage of set theory, real number system, metric spaces, continuous functions, Riemann integration, multiple integrals, more. Wide range of problems. Undergraduate level. Bibliography. 254pp. 5⅜ × 8½. 65038-3 Pa. $7.00

INTRODUCTION TO QUANTUM MECHANICS With Applications to Chemistry, Linus Pauling & E. Bright Wilson, Jr. Classic undergraduate text by Nobel Prize winner applies quantum mechanics to chemical and physical problems. Numerous tables and figures enhance the text. Chapter bibliographies. Appendices. Index. 468pp. 5⅜ × 8½. 64871-0 Pa. $9.95

ASYMPTOTIC EXPANSIONS OF INTEGRALS, Norman Bleistein & Richard A. Handelsman. Best introduction to important field with applications in a variety of scientific disciplines. New preface. Problems. Diagrams. Tables. Bibliography. Index. 448pp. 5⅜ × 8½. 65082-0 Pa. $10.95

MATHEMATICS APPLIED TO CONTINUUM MECHANICS, Lee A. Segel. Analyzes models of fluid flow and solid deformation. For upper-level math, science and engineering students. 608pp. 5⅜ × 8½. 65369-2 Pa. $12.95

ELEMENTS OF REAL ANALYSIS, David A. Sprecher. Classic text covers fundamental concepts, real number system, point sets, functions of a real variable, Fourier series, much more. Over 500 exercises. 352pp. 5⅜ × 8½. 65385-4 Pa. $8.95

PHYSICAL PRINCIPLES OF THE QUANTUM THEORY, Werner Heisenberg. Nobel Laureate discusses quantum theory, uncertainty, wave mechanics, work of Dirac, Schroedinger, Compton, Wilson, Einstein, etc. 184pp. 5⅜ × 8½. 60113-7 Pa. $4.95

INTRODUCTORY REAL ANALYSIS, A.N. Kolmogorov, S.V. Fomin. Translated by Richard A. Silverman. Self-contained, evenly paced introduction to real and functional analysis. Some 350 problems. 403pp. 5⅜ × 8½. 61226-0 Pa. $7.95

PROBLEMS AND SOLUTIONS IN QUANTUM CHEMISTRY AND PHYSICS, Charles S. Johnson, Jr. and Lee G. Pedersen. Unusually varied problems, detailed solutions in coverage of quantum mechanics, wave mechanics, angular momentum, molecular spectroscopy, scattering theory, more. 280 problems plus 139 supplementary exercises. 430pp. 6½ × 9¼. 65236-X Pa. $10.95

THE ELECTROMAGNETIC FIELD, Albert Shadowitz. Comprehensive undergraduate text covers basics of electric and magnetic fields, builds up to electromagnetic theory. Also related topics, including relativity. Over 900 problems. 768pp. 5⅜ × 8¼. 65660-8 Pa. $15.95

FOURIER SERIES, Georgi P. Tolstov. Translated by Richard A. Silverman. A valuable addition to the literature on the subject, moving clearly from subject to subject and theorem to theorem. 107 problems, answers. 336pp. 5⅜ × 8½. 63317-9 Pa. $7.95

THEORY OF ELECTROMAGNETIC WAVE PROPAGATION, Charles Herach Papas. Graduate-level study discusses the Maxwell field equations, radiation from wire antennas, the Doppler effect and more. xiii + 244pp. 5⅜ × 8½. 65678-0 Pa. $6.95

DISTRIBUTION THEORY AND TRANSFORM ANALYSIS: An Introduction to Generalized Functions, with Applications, A.H. Zemanian. Provides basics of distribution theory, describes generalized Fourier and Laplace transformations. Numerous problems. 384pp. 5⅜ × 8½. 65479-6 Pa. $8.95

THE PHYSICS OF WAVES, William C. Elmore and Mark A. Heald. Unique overview of classical wave theory. Acoustics, optics, electromagnetic radiation, more. Ideal as classroom text or for self-study. Problems. 477pp. 5⅜ × 8½. 64926-1 Pa. $10.95

CALCULUS OF VARIATIONS WITH APPLICATIONS, George M. Ewing. Applications-oriented introduction to variational theory develops insight and promotes understanding of specialized books, research papers. Suitable for advanced undergraduate/graduate students as primary, supplementary text. 352pp. 5⅜ × 8½. 64856-7 Pa. $8.50

A TREATISE ON ELECTRICITY AND MAGNETISM, James Clerk Maxwell. Important foundation work of modern physics. Brings to final form Maxwell's theory of electromagnetism and rigorously derives his general equations of field theory. 1,084pp. 5⅜ × 8½. 60636-8, 60637-6 Pa., Two-vol. set $19.00

AN INTRODUCTION TO THE CALCULUS OF VARIATIONS, Charles Fox. Graduate-level text covers variations of an integral, isoperimetrical problems, least action, special relativity, approximations, more. References. 279pp. 5⅜ × 8½. 65499-0 Pa. $6.95

HYDRODYNAMIC AND HYDROMAGNETIC STABILITY, S. Chandrasekhar. Lucid examination of the Rayleigh-Benard problem; clear coverage of the theory of instabilities causing convection. 704pp. 5⅜ × 8¼. 64071-X Pa. $12.95

CALCULUS OF VARIATIONS, Robert Weinstock. Basic introduction covering isoperimetric problems, theory of elasticity, quantum mechanics, electrostatics, etc. Exercises throughout. 326pp. 5⅜ × 8½. 63069-2 Pa. $7.95

DYNAMICS OF FLUIDS IN POROUS MEDIA, Jacob Bear. For advanced students of ground water hydrology, soil mechanics and physics, drainage and irrigation engineering and more. 335 illustrations. Exercises, with answers. 784pp. 6⅛ × 9¼. 65675-6 Pa. $19.95

NUMERICAL METHODS FOR SCIENTISTS AND ENGINEERS, Richard Hamming. Classic text stresses frequency approach in coverage of algorithms, polynomial approximation, Fourier approximation, exponential approximation, other topics. Revised and enlarged 2nd edition. 721pp. 5⅜ × 8½.
65241-6 Pa. $14.95

THEORETICAL SOLID STATE PHYSICS, Vol. I: Perfect Lattices in Equilibrium; Vol. II: Non-Equilibrium and Disorder, William Jones and Norman H. March. Monumental reference work covers fundamental theory of equilibrium properties of perfect crystalline solids, non-equilibrium properties, defects and disordered systems. Appendices. Problems. Preface. Diagrams. Index. Bibliography. Total of 1,301pp. 5⅜ × 8½. Two volumes. Vol. I 65015-4 Pa. $12.95
Vol. II 65016-2 Pa. $12.95

OPTIMIZATION THEORY WITH APPLICATIONS, Donald A. Pierre. Broad-spectrum approach to important topic. Classical theory of minima and maxima, calculus of variations, simplex technique and linear programming, more. Many problems, examples. 640pp. 5⅜ × 8½. 65205-X Pa. $12.95

THE MODERN THEORY OF SOLIDS, Frederick Seitz. First inexpensive edition of classic work on theory of ionic crystals, free-electron theory of metals and semiconductors, molecular binding, much more. 736pp. 5⅜ × 8½.
65482-6 Pa. $14.95

ESSAYS ON THE THEORY OF NUMBERS, Richard Dedekind. Two classic essays by great German mathematician: on the theory of irrational numbers; and on transfinite numbers and properties of natural numbers. 115pp. 5⅜ × 8½.
21010-3 Pa. $4.95

THE FUNCTIONS OF MATHEMATICAL PHYSICS, Harry Hochstadt. Comprehensive treatment of orthogonal polynomials, hypergeometric functions, Hill's equation, much more. Bibliography. Index. 322pp. 5⅜ × 8½. 65214-9 Pa. $8.95

NUMBER THEORY AND ITS HISTORY, Oystein Ore. Unusually clear, accessible introduction covers counting, properties of numbers, prime numbers, much more. Bibliography. 380pp. 5⅜ × 8½. 65620-9 Pa. $8.95

THE VARIATIONAL PRINCIPLES OF MECHANICS, Cornelius Lanzcos. Graduate level coverage of calculus of variations, equations of motion, relativistic mechanics, more. First inexpensive paperbound edition of classic treatise. Index. Bibliography. 418pp. 5⅜ × 8½. 65067-7 Pa. $10.95

MATHEMATICAL TABLES AND FORMULAS, Robert D. Carmichael and Edwin R. Smith. Logarithms, sines, tangents, trig functions, powers, roots, reciprocals, exponential and hyperbolic functions, formulas and theorems. 269pp. 5⅜ × 8½. 60111-0 Pa. $5.95

THEORETICAL PHYSICS, Georg Joos, with Ira M. Freeman. Classic overview covers essential math, mechanics, electromagnetic theory, thermodynamics, quantum mechanics, nuclear physics, other topics. First paperback edition. xxiii + 885pp. 5⅜ × 8½. 65227-0 Pa. $17.95

ROTARY-WING AERODYNAMICS, W.Z. Stepniewski. Clear, concise text covers aerodynamic phenomena of the rotor and offers guidelines for helicopter performance evaluation. Originally prepared for NASA. 537 figures. 640pp. 6⅛ × 9¼.
64647-5 Pa. $14.95

DIFFERENTIAL GEOMETRY, Heinrich W. Guggenheimer. Local differential geometry as an application of advanced calculus and linear algebra. Curvature, transformation groups, surfaces, more. Exercises. 62 figures. 378pp. 5⅜ × 8½.
63433-7 Pa. $7.95

INTRODUCTION TO SPACE DYNAMICS, William Tyrrell Thomson. Comprehensive, classic introduction to space-flight engineering for advanced undergraduate and graduate students. Includes vector algebra, kinematics, transformation of coordinates. Bibliography. Index. 352pp. 5⅜ × 8½. 65113-4 Pa. $8.00

A SURVEY OF MINIMAL SURFACES, Robert Osserman. Up-to-date, in-depth discussion of the field for advanced students. Corrected and enlarged edition covers new developments. Includes numerous problems. 192pp. 5⅜ × 8½.
64998-9 Pa. $8.00

ANALYTICAL MECHANICS OF GEARS, Earle Buckingham. Indispensable reference for modern gear manufacture covers conjugate gear-tooth action, gear-tooth profiles of various gears, many other topics. 263 figures. 102 tables. 546pp. 5⅜ × 8½.
65712-4 Pa. $11.95

SET THEORY AND LOGIC, Robert R. Stoll. Lucid introduction to unified theory of mathematical concepts. Set theory and logic seen as tools for conceptual understanding of real number system. 496pp. 5⅜ × 8¼. 63829-4 Pa. $8.95

A HISTORY OF MECHANICS, René Dugas. Monumental study of mechanical principles from antiquity to quantum mechanics. Contributions of ancient Greeks, Galileo, Leonardo, Kepler, Lagrange, many others. 671pp. 5⅜ × 8½.
65632-2 Pa. $14.95

FAMOUS PROBLEMS OF GEOMETRY AND HOW TO SOLVE THEM, Benjamin Bold. Squaring the circle, trisecting the angle, duplicating the cube: learn their history, why they are impossible to solve, then solve them yourself. 128pp. 5⅜ × 8½.
24297-8 Pa. $3.95

MECHANICAL VIBRATIONS, J.P. Den Hartog. Classic textbook offers lucid explanations and illustrative models, applying theories of vibrations to a variety of practical industrial engineering problems. Numerous figures. 233 problems, solutions. Appendix. Index. Preface. 436pp. 5⅜ × 8½. 64785-4 Pa. $8.95

CURVATURE AND HOMOLOGY, Samuel I. Goldberg. Thorough treatment of specialized branch of differential geometry. Covers Riemannian manifolds, topology of differentiable manifolds, compact Lie groups, other topics. Exercises. 315pp. 5⅜ × 8½.
64314-X Pa. $6.95

HISTORY OF STRENGTH OF MATERIALS, Stephen P. Timoshenko. Excellent historical survey of the strength of materials with many references to the theories of elasticity and structure. 245 figures. 452pp. 5⅜ × 8½. 61187-6 Pa. $9.95

THE FOUR-COLOR PROBLEM: Assaults and Conquest, Thomas L. Saaty and Paul G. Kainen. Engrossing, comprehensive account of the century-old combinatorial topological problem, its history and solution. Bibliographies. Index. 110 figures. 228pp. 5⅜ × 8½. 65092-8 Pa. $6.00

CATALYSIS IN CHEMISTRY AND ENZYMOLOGY, William P. Jencks. Exceptionally clear coverage of mechanisms for catalysis, forces in aqueous solution, carbonyl- and acyl-group reactions, practical kinetics, more. 864pp. 5⅜ × 8½. 65460-5 Pa. $18.95

PROBABILITY: An Introduction, Samuel Goldberg. Excellent basic text covers set theory, probability theory for finite sample spaces, binomial theorem, much more. 360 problems. Bibliographies. 322pp. 5⅜ × 8½. 65252-1 Pa. $7.95

LIGHTNING, Martin A. Uman. Revised, updated edition of classic work on the physics of lightning. Phenomena, terminology, measurement, photography, spectroscopy, thunder, more. Reviews recent research. Bibliography. Indices. 320pp. 5⅜ × 8¼. 64575-4 Pa. $7.95

PROBABILITY THEORY: A Concise Course, Y.A. Rozanov. Highly readable, self-contained introduction covers combination of events, dependent events, Bernoulli trials, etc. Translation by Richard Silverman. 148pp. 5⅜ × 8¼. 63544-9 Pa. $4.50

THE CEASELESS WIND: An Introduction to the Theory of Atmospheric Motion, John A. Dutton. Acclaimed text integrates disciplines of mathematics and physics for full understanding of dynamics of atmospheric motion. Over 400 problems. Index. 97 illustrations. 640pp. 6 × 9. 65096-0 Pa. $16.95

STATISTICS MANUAL, Edwin L. Crow, et al. Comprehensive, practical collection of classical and modern methods prepared by U.S. Naval Ordnance Test Station. Stress on use. Basics of statistics assumed. 288pp. 5⅜ × 8½. 60599-X Pa. $6.00

WIND WAVES: Their Generation and Propagation on the Ocean Surface, Blair Kinsman. Classic of oceanography offers detailed discussion of stochastic processes and power spectral analysis that revolutionized ocean wave theory. Rigorous, lucid. 676pp. 5⅜ × 8½. 64652-1 Pa. $14.95

STATISTICAL METHOD FROM THE VIEWPOINT OF QUALITY CONTROL, Walter A. Shewhart. Important text explains regulation of variables, uses of statistical control to achieve quality control in industry, agriculture, other areas. 192pp. 5⅜ × 8½. 65232-7 Pa. $6.00

THE INTERPRETATION OF GEOLOGICAL PHASE DIAGRAMS, Ernest G. Ehlers. Clear, concise text emphasizes diagrams of systems under fluid or containing pressure; also coverage of complex binary systems, hydrothermal melting, more. 288pp. 6½ × 9¼. 65389-7 Pa. $8.95

STATISTICAL ADJUSTMENT OF DATA, W. Edwards Deming. Introduction to basic concepts of statistics, curve fitting, least squares solution, conditions without parameter, conditions containing parameters. 26 exercises worked out. 271pp. 5⅜ × 8½. 64685-8 Pa. $7.95